AF808993

CliffsNotes® Algebra II Common Core Quick Review

CliffsNotes® Algebra II Common Core Quick Review

By Wendy Taub-Hoglund, M.S.

About the Author

Wendy Taub-Hoglund, M.S., is a highly regarded author and educator who has enhanced the field of mathematics education through instructing, mentoring, and writing curriculum and assessments for the California public schools.

She taught middle and high school mathematics for the Los Angeles Unified School District and the Beverly Hills Unified School District for over 39 years. She currently teaches mathematics in the Los Angeles Community College and South Lake Tahoe Community College districts.

Dedication

Dedicated to all those students who have asked over many years of teaching "when will I ever need to know this?"

Editorial

Executive Editor: Greg Tubach
Senior Editor: Christina Stambaugh
Production Editor: Erika West
Copy Editor: Lynn Northrup
Technical Editors: Mary Jane Sterling and Tom Page
Proofreader: Susan Moritz
Indexer: Potomac Indexing, LLC

CliffsNotes® Algebra II Common Core Quick Review

Library of Congress Control Number: 2016950678 ISBN: 978-1-957671-14-7

Printed in the United States of America
DOC 10 9 8 7 6 5 4 3 2 1

Table of Contents

INTRODUCTION

CliffsNotes Algebra II Common Core Quick Review is a comprehensive study guide to many topics in a second course of algebra, including information on subjects ranging from linear equations in one, two, and three variables, to inequalities, complex numbers, conic sections, quadratic equations, logarithms, and trigonometry. Whether you are looking for an in-depth reference to subject matter of the course or occasional reinforcement of one or more aspects of algebra, it is contained in this book.

For the purpose of this review, your knowledge of the following fundamental ideas is assumed:

- Natural numbers
- Whole numbers
- Integers
- Rational numbers
- Irrational numbers
- Real numbers
- Rules for operations with positive and negative numbers
- Properties of exponents
- Grouping symbols
- Order of operations
- Scientific notation
- Absolute value

If you feel that you are weak in any of these topics, refer to *CliffsNotes Algebra I Common Core Quick Review* or *CliffsNotes Basic Math and Pre-Algebra Quick Review.*

Common Core State Standards for Mathematics (CCSSM)

CliffsNotes Algebra II Common Core Quick Review topics are aligned to the Common Core State Standards for Mathematics (CCSSM)—the skills you should know and be able to perform in Algebra II. The concepts presented in this book are closely interrelated to a broader set of Common Core Mathematics conceptual categories:

- Numbers and quantity
- Algebra
- Functions
- Modeling
- Geometry
- Statistics and probability

Our goal is to help you become an advanced "mathematical thinker" so that you can make logical connections between the topics presented in this book and in each of the Common Core domains. *CliffsNotes Algebra II Common Core Quick Review* gives you procedural approaches to increase your knowledge, fluency, and skills in higher-level mathematics and real-world algebraic applications.

Common Core Mathematics falls into two interconnecting conceptual categories: those that address knowing content standard skills and those that address practicing and applying standard processes. Both categories are aimed at readying students to apply their mathematical knowledge to higher-level mathematics and career mathematics. The examples presented throughout this book exemplify both the content and practice of Common Core Mathematics Standards.

Connecting to Common Core Mathematics

Use the following guidelines in the practice of the Standards of Common Core Mathematics algebraic topics:

- Make sense of problems, persevere in solving problems, and continue to monitor and evaluate your progress.

- Reason abstractly and quantitatively to describe relationships between numbers and variables.
- Construct viable arguments and analyze your reasoning as you deconstruct, organize, and make sense of algebraic problems.
- Justify your plausible conclusions and clarify your explanations of algebraic problems.
- Model with mathematics by mapping, drawing, graphing, and using visual diagrams to validate your conclusion in the context of an algebraic scenario.
- Use appropriate electronic tools (i.e., calculators, software, or any other technology) to strengthen your understanding of algebraic concepts.
- Attend to precision as you carefully check your own reasoning to find the correct solutions.
- Look for and make use of structural patterns in algebraic problems while paying attention to groupings, properties, and order of operations.
- Look for and express algebraic reasoning problems regularly to improve your algebraic proficiency.

Why You Need This Book

Can you answer "yes" to any of the following questions?

- Do you need to review the fundamentals of Algebra II fast?
- Do you need a course supplement to Algebra II?
- Do you need to prepare for your Algebra II test?
- Do you need a concise, comprehensive reference for Algebra II?
- Do you need a no-nonsense approach to Algebra II that gets you the results you need?
- Do you need practice with real-life applications of Algebra II topics?

If so, then *CliffsNotes Algebra II Common Core Quick Review* is for you!

How to Use This Book

You're in charge here. You get to decide how to use this book. You can read it cover to cover, or just look for the information that you want and then put the book back on the shelf for later use. Here are a few of the recommended ways to search for information about a particular Algebra II topic:

- Look for areas of interest in the book's table of contents or use the index to find specific topics.
- Flip through the book, looking for subject areas by heading.
- Get a glimpse of what you'll gain from a chapter by reading through the "Chapter Check-In" and "Common Core Standard" references at the beginning of each chapter.
- Use the "Chapter Check-Out" at the end of each chapter to gauge your grasp of the important information you need to know.
- Test your knowledge more completely in the "Review Questions" (pp. 321–328).
- Look in the glossary (pp. 329–336) for important terms and definitions. If a word is boldfaced in the text, you can find a more complete definition in the book's glossary.

Hundreds of Practice Questions Online!

Go to CliffsNotes.com for hundreds of additional Algebra II Common Core practice questions to help you prepare for your next quiz or test. The questions are organized by this book's chapter sections, so it is easy to use the book and then quiz yourself online to make sure you know the subject. Visit CliffsNotes.com to test yourself anytime and find other free homework help.

Chapter 1

LINEAR SENTENCES IN ONE VARIABLE

Chapter Check-In

- ❑ Defining and solving linear equations—including special cases
- ❑ Solving problems by using formulas
- ❑ Working with absolute value equations
- ❑ Understanding linear inequalities
- ❑ Dealing with compound inequalities
- ❑ Solving absolute value inequalities

Common Core Standard: Seeing the Structure in Expressions

Linear equations, inequalities, and compound inequalities in one variable. Create equations that describe numbers or relationships (A.CED.1-4). Interpret the structure of expressions and write expressions in equivalent forms (A.SSE.2-4). Understand solving equations as a process of reasoning and explain the reasoning (A.REI.1).

Linear sentences in one variable may be equations or inequalities. What they have in common is that the variable has an exponent of 1, which is understood and never written (except for teaching purposes). Equations can be represented on a graph in the form of a straight line. Inequalities can be represented on a graph in the form of a straight line and a shaded region (or a half plane). Equations and inequalities are covered in chapters 2 and 3. This chapter examines different ways of writing and solving linear sentences.

Linear Equations

An **equation** is a statement that implies equality between two mathematical expressions—one on the left side of the equal sign (=) and one on the right side of the equal sign.

A **linear equation in one variable** is an equation with an exponent 1 on the variable. These are also known as **first-degree equations** because the highest exponent on the variable is 1. All linear equations eventually can be written in the form $ax + b = c$, where a, b, and c are real numbers and $a \neq 0$. It is assumed that you are familiar with the addition and multiplication properties of equality.

- *Addition property of equality.* The addition property of equality states that the same term or quantity can be added to both sides of an equation without changing the condition of equality.

  For example, if a, b, and c are real numbers and $a = b$, then $a + c = b + c$.

- *Multiplication property of equality.* The multiplication property of equality states that both sides of an equation can be multiplied by the same quantity without changing the condition of equality.

  For example, if a, b, and c are real numbers and $a = b$, then $ac = bc$.

The goal in solving linear equations is to isolate the variable on either side of the equation by first using the addition property of equality and then using the multiplication property of equality to change the coefficient of the variable to 1.

Example 1: Solve for x: $6(2x - 5) = 4(8x + 7)$.

$$6(2x-5)=4(8x+7)$$
$$12x-30=32x+28$$

To isolate the x on either side of the equation, you can either add $-12x$ to both sides or add $-32x$ to both sides.

$$\begin{aligned} 12x-30&=32x+28 \\ \underline{-12x} \qquad &\quad \underline{-12x} \\ -30&=20x+28 \end{aligned}$$

Isolate the $20x$.

$$\begin{aligned} \underline{-28} \qquad &\quad \underline{-28} \\ -58&=20x \end{aligned}$$

Multiply each side by $\frac{1}{20}$ (or divide each side by 20).

$$\frac{1}{20}(-58) = \frac{1}{20}(20x)$$
$$-\frac{29}{10} = x$$

The solution is $-\frac{29}{10}$. The list of all possible values for x form a **solution set,** written as braces around a set of numbers: $\left\{-\frac{29}{10}\right\}$. You can check this solution by replacing x with $-\frac{29}{10}$ in the original equation.

Example 2: Solve for x: $\frac{2x-5}{4} - \frac{x}{3} = 2 - \frac{x+4}{6}$.

This equation will be made simpler to solve by first clearing the equation of fractions. To do this, find the least common denominator (LCD) for all the denominators in the equation and multiply both sides of the equation by this value, using the distributive property.

$$\frac{2x-5}{4} - \frac{x}{3} = 2 - \frac{x+4}{6}$$
$$\text{LCD} = 12$$
$$12\left(\frac{2x-5}{4} - \frac{x}{3}\right) = 12\left(2 - \frac{x+4}{6}\right)$$
$$\overset{3}{\cancel{12}}\left(\frac{2x-5}{\underset{1}{\cancel{4}}}\right) - \overset{4}{\cancel{12}}\left(\frac{x}{\underset{1}{\cancel{3}}}\right) = 12(2) - \overset{2}{\cancel{12}}\left(\frac{x+4}{\underset{1}{\cancel{6}}}\right)$$
$$6x - 15 - 4x = 24 - 2x - 8$$

Note that the -2 is distributed over *both* the x and the 4. Simplify both sides by combining like terms.

$$2x-15=16-2x$$

Get the variable on one side. $\quad \underline{+2x} \qquad \underline{+2x}$

$$4x-15=16$$

Isolate the variable. $\quad \underline{+15} \ \underline{+15}$

$$4x \quad =31$$

Multiply each side by $\frac{1}{4}$ (or divide each side by 4).

$$\frac{1}{4}(4x)=\frac{1}{4}(31)$$

$$x=\frac{31}{4}$$

You can check this for yourself. The solution set is $\left\{\frac{31}{4}\right\}$.

An **identity equation** is true for all values of a variable regardless of the value plugged in for the variable. The solution to an identity equation is always a real number. Some equations are *never true* for any value of the variable. This type of equation has no solution. The symbol for the solution set of an equation that is never true is "∅."

Example 3: Solve for x: $4-4x=2(2-2x)$.

$$4-4x=2(2-2x)$$

$4-4x=4-4x$ Remove the parentheses.

$\underline{\quad +4x \qquad +4x \quad}$ Add $4x$ to each side.

$$4=4$$

This equation is true for all values of x.

Example 4: Solve for x: $5(x-3)=2x+3(x+1)$.

$$5x-15=2x+3x+3$$

$$5x-15=5x+3$$

$$-15\neq 3$$

There are no values of x that will make this equation true, ∅.

Formulas

A **formula** is an equation that describes a relationship between unknown values. Many Common Core Mathematics problems are easily solved and analyzed if the correct formula is known. A formula is similar to a recipe because it offers written rules to tell you how to find one quantity when the other quantities upon which it depends are given. To use formulas to solve problems, perform the following steps:

1. Identify the appropriate formula.
2. Replace the variables in the formula with their known values.
3. Solve the formula for the remaining variable(s).
4. Check the solution in the original problem.
5. State the solution.

Example 1: Find the length of the following rectangle with a perimeter of 48 inches and a width of 8 inches.

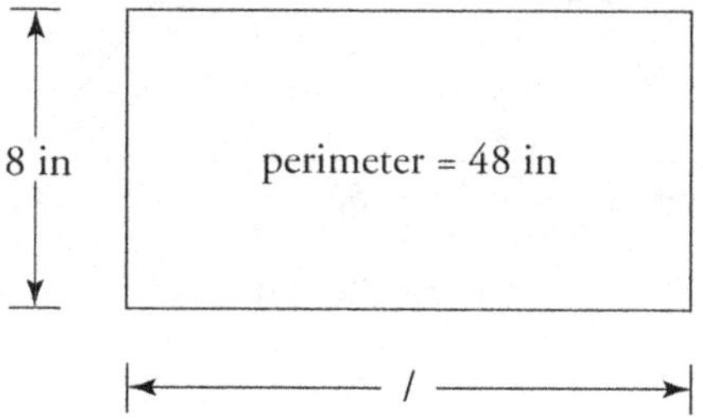

An appropriate formula is $p = 2l + 2w$, where p represents the perimeter, l the length, and w the width of the rectangle. Replace each variable with its known value.

$$p = 2l + 2w$$

$$48 = 2l + 2(8)$$

Now, solve for the remaining variable.

$$48 = 2l + 16$$

Add -16 to both sides.

$$32 = 2l$$

Multiply each side by $\frac{1}{2}$ (or divide each side by 2).

$$16 = l$$

Check with the original problem.

$$48 = 2l + 2(8)$$
$$48 \stackrel{?}{=} 2(16) + 2(8)$$
$$48 \stackrel{?}{=} 32 + 16$$
$$48 = 48 \checkmark$$

The length of the rectangle is 16 inches.

Example 2: As shown in the following figure, an area rug is in the shape of a trapezoid. Its area is 38 square feet, its height is 4 feet, and one of its bases is 7 feet long. Find the length of the other base.

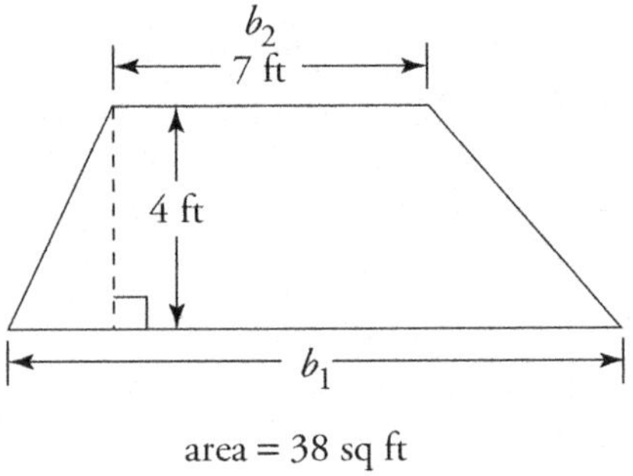

$$A_{\text{trapezoid}} = \frac{1}{2}h(b_1 + b_2)$$
$$38 = \frac{1}{2}(4)(7 + b_2)$$
$$38 = 14 + 2b_2$$
$$24 = 2b_2$$
$$12 = b_2$$

The other base has a length of 12 feet.

Many times, a formula needs to be solved for one of its variables in terms of the other variables. To do this, simply perform the steps for solving equations as before but, this time, isolate the specified variable.

Example 3: $rt = d$: (rate × time = distance). Solve for r.

$$rt = d$$
$$\frac{1}{t}(rt) = \frac{1}{t}(d)$$
$$r = \frac{d}{t}$$

Example 4: $A = P + PRT$: (*A*ccumulated interest = *P*rincipal + *P*rincipal × *R*ate × *T*ime). Solve for T.

$$A = P + PRT$$
$$A - P = PRT$$
$$\frac{1}{PR}(A-P) = \frac{1}{PR}(PRT)$$
$$\frac{A-P}{PR} = T$$

Absolute Value Equations

Common Core Mathematics asks you to solve equations that may include absolute values. Recall that the absolute value of a number represents the distance from zero to that number on the number line. The equation $|x| = 3$ is translated as "x is 3 units from zero on the number line." Notice that, on the number line that follows, two different numbers are each 3 units away from zero, namely, 3 and -3.

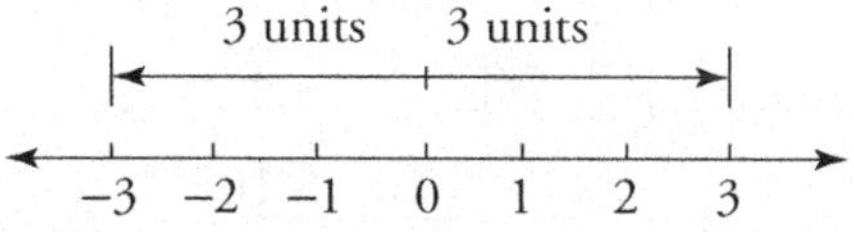

The solution set of the equation $|x| = 3$ is $\{3, -3\}$ because $|3| = 3$ and $|-3| = 3$.

Example 1: Solve for x: $|4x - 2| = 8$.

This translates to "$4x - 2$ is 8 units from zero on the number line." There are positive and negative solutions, as shown on the following graph.

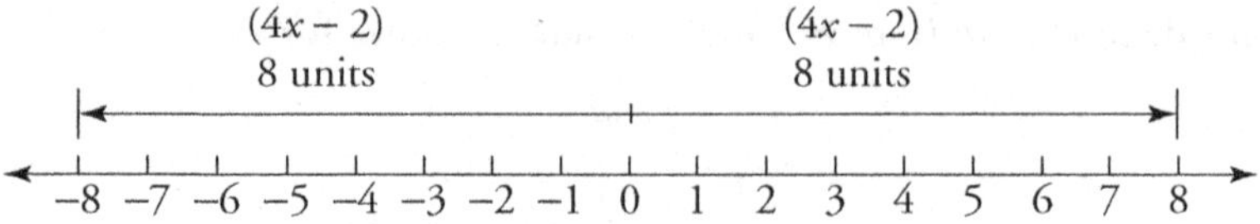

$$4x-2=-8 \quad \text{or} \quad 4x-2=8$$

$$4x=-6 \qquad 4x=10$$

$$x=-\frac{3}{2} \qquad x=\frac{5}{2}$$

Check the solutions.

$$|4x-2|=8 \quad \text{or} \quad |4x-2|=8$$

$$\left|4\left(-\frac{3}{2}\right)-2\right|\stackrel{?}{=}8 \qquad \left|4\left(\frac{5}{2}\right)-2\right|\stackrel{?}{=}8$$

$$|-6-2|\stackrel{?}{=}8 \qquad |10-2|\stackrel{?}{=}8$$

$$|-8|\stackrel{?}{=}8 \qquad |8|\stackrel{?}{=}8$$

$$8=8\checkmark \qquad 8=8\checkmark$$

These are both true statements. The solution set is $\left\{-\frac{3}{2}, \frac{5}{2}\right\}$.

Example 2: Solve for x: $\left|\frac{3x}{2}+2\right|+10=21$.

To solve this type of absolute value equation, first isolate the absolute value expression.

$$\left|\frac{3x}{2}+2\right|+10= \ 21$$

$$\underline{-10} \quad \underline{-10}$$

$$\left|\frac{3x}{2}+2\right|= \ 11$$

Now, translate the absolute value equation: "$\frac{3x}{2}+2$ is 11 units from zero on the number line."

$$\frac{3x}{2}+2=-11 \quad \text{or} \quad \frac{3x}{2}+2=11$$

$$2\left(\frac{3x}{2}+2\right)=2(-11) \qquad 2\left(\frac{3x}{2}+2\right)=2(11)$$

$$3x+4=-22 \qquad 3x+4=22$$

$$3x=-26 \qquad 3x=18$$

$$x=-\frac{26}{3} \qquad x=6$$

The solution set is $\left\{-\frac{26}{3}, 6\right\}$.

Example 3: Solve for x: $|x| = -2$.

This problem has no solutions because the translation does not make mathematical sense. Distance is not measured in negative values.

Example 4: Solve for x: $|2x - 3| = |3x + 7|$.

This type of sentence will be true if either

- The expressions *inside* the absolute value symbols are exactly the same (that is, they are equal), or
- The expressions *inside* the absolute value symbols are opposites of each other.

$$2x-3=3x+7 \quad \text{or} \quad 2x-3=-(3x+7)$$

$$-3=x+7 \qquad 2x-3=-3x-7$$

$$-10=x \qquad 5x-3=-7$$

$$5x=-4$$

$$x=-\frac{4}{5}$$

The solution set is $\left\{-10, -\frac{4}{5}\right\}$.

Example 5: Solve for x: $|x - 2| = |7 - x|$.

$$\begin{aligned} x-2&=7-x \\ 2x-2&=7 \\ 2x&=9 \\ x&=\frac{9}{2} \end{aligned} \qquad \text{or} \qquad \begin{aligned} x-2&=-(7-x) \\ x-2&=-7+x \\ -2&=-7 \end{aligned}$$

The sentence $-2 = -7$ is never true, so it gives no solution. So the only possible solution is $x = \frac{9}{2}$.

Check the solution.

$$\begin{aligned} |x-2| &= |7-x| \\ \left|\frac{9}{2}-2\right| &\stackrel{?}{=} \left|7-\frac{9}{2}\right| \\ \left|\frac{9}{2}-\frac{4}{2}\right| &\stackrel{?}{=} \left|\frac{14}{2}-\frac{9}{2}\right| \\ \left|\frac{5}{2}\right| &\stackrel{?}{=} \left|\frac{5}{2}\right| \\ \frac{5}{2} &= \frac{5}{2} \checkmark \end{aligned}$$

The solution set is $\left\{\frac{9}{2}\right\}$.

Example 6: The value 135 can be separated into three numbers. The second number is twice the first number, and the third number is ten more than the second number. Find all three numbers.

x = first number
$2x$ = second number
$2x + 10$ = third number

These three numbers, when combined (or added), should equal 135, giving the equation:

$$\begin{aligned} x+2x+(2x+10)&=135 \\ 5x+10&=135 \\ 5x&=125 \\ x&=25 \end{aligned}$$

The numbers are 25, 50, and 60.

Example 7: One acute angle of a right triangle measures 60 degrees more than twice the other acute angle. Find the measures of the acute angles in this right triangle.

x = measure of angle 1
$2x + 60$ = measure of angle 2

$$x+(2x+60)+90=180$$
$$3x+150=180$$
$$3x=30$$
$$x=10$$

The acute angles measure 10 degrees and 80 degrees.

Example 8: An aunt is twice as old as her niece. In 4 years from now or $2x + 4$, the aunt's age will be three times the age that the niece was 6 years ago or $3(x - 6)$. How old is each family member now?

niece's age now $= x$
aunt's age now $= 2x$

$$2x+4=3(x-6)$$
$$2x+4=3x-18$$
$$-x=-22$$
$$x=22$$

The aunt is 44 years old, and the niece is 22 years old.

Example 9: The formula for the surface area of a sphere (S), where r is the radius, is $S = 4\pi r^2$. Find the radius of a sphere whose surface area is 180 square meters. Use 3.14 for π; round your answer to the nearest tenth.

$$180=4\pi r^2$$
$$180=4(3.14)r^2$$
$$180=12.56r^2$$
$$14.33=r^2$$
$$3.79\approx r$$

The radius of the sphere is approximately 3.8.

Example 10: How much cranberry juice costing \$0.90 per gallon should be mixed with lemonade costing \$0.60 per gallon to make 100 gallons of a mixture worth \$0.75 per gallon?

x = amount of cranberry juice
$100 - x$ = amount of lemonade

$$\begin{aligned} 0.90x + 0.60(100 - x) &= 0.75(100) \\ 0.90x + 60 - 0.60x &= 75 \\ 0.30x + 60 &= 75 \\ 0.30x &= 15 \\ x &= 50 \end{aligned}$$

To make 100 gallons of a mixture worth \$0.75 per gallon, you'll need 50 gallons of cranberry juice and 50 gallons of lemonade.

Linear Inequalities

An **inequality** is a mathematical sentence using a symbol other than the equal sign (=). The most common inequality symbols are $<$, $\leq$, $>$, and $\geq$. To solve an inequality sentence, use exactly the same procedure that you would if it were an equation, with the following exception: When multiplying (or dividing) both sides of an inequality by a negative number, the direction of the inequality sign reverses. This is called the *negative multiplication property of inequality.*

> **Negative Multiplication Property of Inequality**
>
> If a, b, and c are real numbers and c is negative and $a < b$, then $ac > bc$. Or if $a > b$, then $ac < bc$.

Example 1: Solve for x: $3x - 7 > 20$.

$$\begin{aligned} 3x - 7 &> 20 \\ 3x &> 27 \\ x &> 9 \end{aligned}$$

To check the solution, first see whether $x = 9$ makes the equation $3x - 7 = 20$ true. Even though 9 isn't a solution, it's a critical number and is important to finding the solution to the inequality.

$$3x - 7 = 20$$
$$3(9) - 7 \stackrel{?}{=} 20$$
$$27 - 7 \stackrel{?}{=} 20$$
$$20 = 20 \checkmark$$

Now, choose a number greater than 9 (try 10, for example), and see whether that makes the original inequality true.

$$3x - 7 > 20$$
$$3(10) - 7 \stackrel{?}{>} 20$$
$$30 - 7 \stackrel{?}{>} 20$$
$$23 > 20 \checkmark$$

This is a true statement. Since it is impossible to list all the numbers that are greater than 9, use set-builder notation to show the solution set.

$$\{x|x > 9\}$$

This is read as "the set of all values of x such that x is greater than 9." Many times, the solutions to inequalities are graphed to illustrate the answers. The graph of $\{x|x > 9\}$ is shown here. Note that 9 is *not* included, as indicated by the open circle. Interval notation is also commonly used. The solution is $(9, \infty)$ using interval notation.

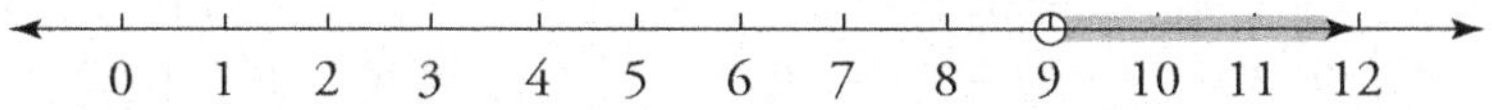

Example 2: Solve for x: $-\frac{5}{8}x + \frac{2}{3} > \frac{3}{4}x - \frac{7}{12}$.

The LCD for the denominators in this inequality is 24. Multiply both sides of the inequality by 24 as you would have if this had been an equation.

$$24\left(-\frac{5}{8}x + \frac{2}{3}\right) > 24\left(\frac{3}{4}x - \frac{7}{12}\right)$$
$$-15x + 16 > 18x - 14$$

At this point, you can isolate x on either side of the inequality.

Isolating x on the left side,	Isolating x on the right side,
$-33x+16>-14$	$16>33x-14$
$-33x>-30$	$30>33x$
$x<\frac{10}{11}$	$\frac{10}{11}>x$

In the final step on the left, the direction is switched because both sides are multiplied or divided by a negative number. Both methods produce the final result that reads: x is a number less than $\frac{10}{11}$.

The solution set is expressed as $\left\{x \mid x<\frac{10}{11}\right\}$.

The interval notation is expressed as $\left(-\infty, \frac{10}{11}\right)$.

The graph of this solution set is shown here. Note that $\frac{10}{11}$ is *not* included, as indicated by the open circle.

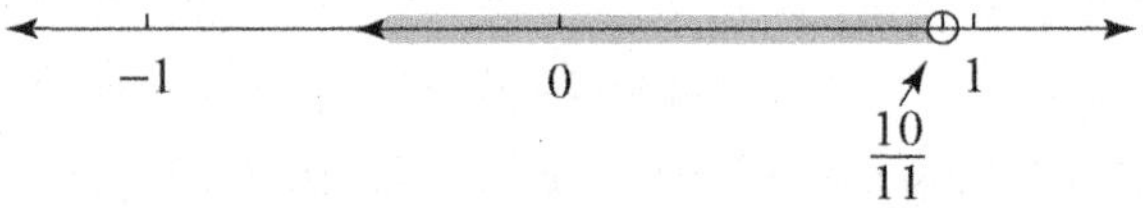

Compound Inequalities

A **compound inequality** is a mathematical sentence with two inequality statements joined either by the word "or" or by the word "and." The use of "and" indicates that both statements of the compound sentence are true at the same time. It is the overlap or *intersection* of the solution sets for the individual statements. The use of "or" indicates that, as long as either statement is true, the entire compound sentence is true. It is the combination or *union* of the solution sets for the individual statements. A compound inequality that uses the word "and" is known as a **conjunction.** Although "and" and "or" are parts of speech known as conjunctions, the mathematical conjunction has a different meaning from the grammatical one. To prove the point, the conjunction (part of speech) "or"—when used in a compound inequality—forms what is known as a **disjunction.** Just remember "con" means "with another," and "dis" means "one OR the other."

Example 1: Solve for x: $3x + 2 < 14$ and $2x - 5 > -11$.

Solve each inequality separately. The joining word "and" indicates that the overlap or intersection is the desired result.

$$\begin{array}{rcr} 3x+2<14 & \text{and} & 2x-5>-11 \\ 3x<12 & & 2x>-6 \\ x<4 & & x>-3 \end{array}$$

$x < 4$ indicates all the numbers to the left of 4, and $x > -3$ indicates all the numbers to the right of -3. The intersection of these two graphs is all the numbers between -3 and 4. The solution set is

$$\{x|x > -3 \text{ and } x < 4\}$$

Another way this solution set could be expressed is

$$\{x|-3 < x < 4\}$$

The interval notation is written $(-3, 4)$.

When a compound inequality is written without the expressed word "and" or "or," it is understood to automatically be the word "and." Reading $\{x|-3 < x < 4\}$ from the "x" position, you say (reading to the left), "x is greater than -3 and (reading to the right) x is less than 4." The graph of the solution set is shown here.

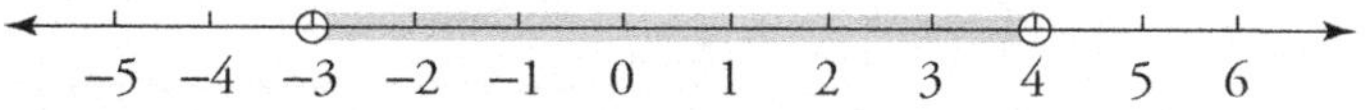

Example 2: Solve for x: $2x + 7 < -11$ or $-3x - 2 < 13$.

Solve each inequality separately. The joining word is "or," so you need to combine the answers; that is, find the union of the solution sets of each inequality sentence.

$$\begin{array}{rcr} 2x+7<-11 & \text{or} & -3x-2<13 \\ 2x<-18 & & -3x<15 \\ x<-9 & & x>-5 \end{array}$$

Remember, as in the last step on the right, to switch the inequality when multiplying by a negative.

$x < -9$ indicates all the numbers to the left of -9; $x > -5$ indicates all the numbers to the right of -5. The solution set is written as $\{x|x < -9 \text{ or } x > -5\}$. The interval notation is written as $(-\infty, -9) \cup (-5, \infty)$.

Here is the graph of the solution set.

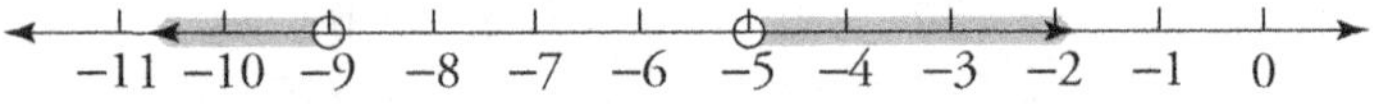

Example 3: Solve for x: $-12 \leq 2x + 6 \leq 8$.

Since this compound inequality has no connecting word written, it is understood to be "and." It is translated into the following compound sentence.

$$-12 \leq 2x+6 \quad \text{and} \quad 2x+6 \leq 8$$
$$-18 \leq 2x \qquad\qquad 2x \leq 2$$
$$-9 \leq x \qquad\qquad x \leq 1$$

$-9 \leq x$ indicates all the numbers to the right of and including -9, and $x \leq 1$ indicates all the numbers to the left of and including 1. The intersection of these graphs is the numbers between -9 and 1, including -9 and 1.

The solution set can be written as $\{x|x \geq -9 \text{ and } x \leq 1\}$ or as $\{x|-9 \leq x \leq 1\}$. The interval notation can be written as $[-9, 1]$.

The graph of the solution set is shown here. Note that -9 and 1 are part of the solution, as indicated by the solid circles.

Example 4: Solve for x: $3x - 2 > -8$ or $2x + 1 < 9$.

$$3x-2>-8 \quad \text{or} \quad 2x+1<9$$
$$3x>-6 \qquad\qquad 2x<8$$
$$x>-2 \qquad\qquad x<4$$

$x > -2$ indicates all the numbers to the right of -2; $x < 4$ indicates all the numbers to the left of 4. The union of these graphs is the entire number line. That is, the solution set is all real numbers. The interval notation is written as $(-\infty, \infty)$. The graph of the solution set is the entire number line, as shown here. The arrowheads indicate infinity.

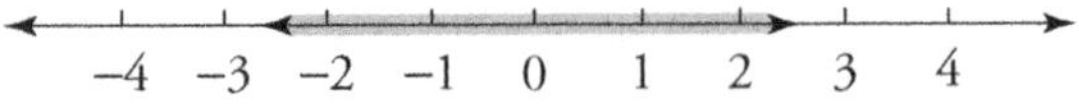

Example 5: Solve for x: $4x - 2 < 10$ and $3x + 1 > 22$.

$$\begin{array}{rcr} 4x-2<10 & \text{and} & 3x+1>22 \\ 4x<12 & & 3x>21 \\ x<3 & & x>7 \end{array}$$

$x < 3$ indicates all the numbers to the left of 3, and $x > 7$ indicates all the numbers to the right of 7. The intersection of these graphs contains no numbers. That is, the solution set is the empty set, ∅.

Absolute Value Inequalities

Remember, absolute value means distance from zero on a number line. $|x| < 4$ means x is a number that is less than 4 units from zero on a number line, as shown in the following graph.

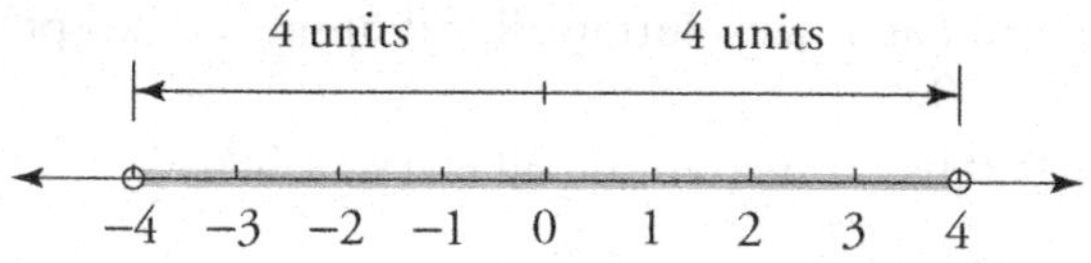

The solutions are the numbers to the right of -4 *and* to the left of 4 and could be indicated as

$$\{x \mid x > -4 \text{ and } x < 4\} \qquad \text{or} \qquad \{x \mid -4 < x < 4\}$$

The interval notation is indicated as $(-4, 4)$.

$|x| > 4$ means x is a number that is more than 4 units from zero on a number line, as shown in the following graph.

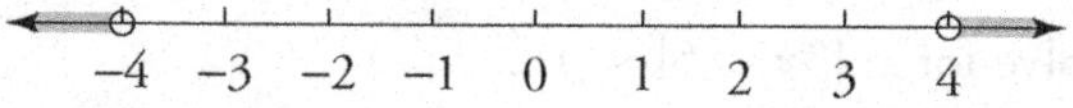

The solutions are the numbers to the left of -4 *or* to the right of 4 and are indicated as $\{x|x < -4 \text{ or } x > 4\}$. The interval notation is indicated as $(-\infty, -4) \cup (4, \infty)$.

$|x| < 0$ has no solutions. $|x| > 0$ has as its solution all real numbers except 0. $|x| > -1$ has as its solution all real numbers because after taking the absolute value of any number, that answer is either zero or positive and will always be greater than -1.

The following is a general approach for solving absolute value inequalities of the forms

$$|ax + b| < c \text{ or } |ax + b| > c$$
$$|ax + b| \leq c \text{ or } |ax + b| \geq c$$

- If c is negative,

  $|ax + b| < c$ has no solutions.

  $|ax + b| \leq c$ has no solutions.

  $|ax + b| > c$ has as its solution all real numbers.

  $|ax + b| \geq c$ has as its solution all real numbers.

- If $c = 0$,

  $|ax + b| < 0$ has no solutions.

  $|ax + b| \leq 0$ has as its solution the solution to $ax + b = 0$.

  $|ax + b| > 0$ has as its solution all real numbers, except the solution to $ax + b = 0$.

  $|ax + b| \geq 0$ has as its solution all real numbers.

- If c is positive,

  $|ax + b| < c$ has solutions that solve $ax + b > -c$ and $ax + b < c$.

  $-c < ax + b < c$.

  That is:

  $|ax + b| \leq c$ has solutions that solve $-c \leq ax + b \leq c$, which becomes "and."

  $|ax + b| \geq c$ has solutions that solve $ax + b \leq -c$ or $ax + b \geq c$.

Example 1: Solve for x: $|3x - 5| < 12$.

$$3x - 5 > -12 \quad \text{and} \quad 3x - 5 < 12$$
$$3x > -7 \qquad\qquad 3x < 17$$
$$x > -\frac{7}{3} \qquad\qquad x < \frac{17}{3}$$

The solution set is

$$\left\{x \mid x > -\frac{7}{3} \text{ and } x < \frac{17}{3}\right\} \qquad \left\{x \mid -\frac{7}{3} < x < \frac{17}{3}\right\}$$

The interval notation is $\left(-\frac{7}{3}, \frac{17}{3}\right)$.

The graph of the solution set is shown in the following graph: x is greater than $-\frac{7}{3}$ and less than $\frac{17}{3}$.

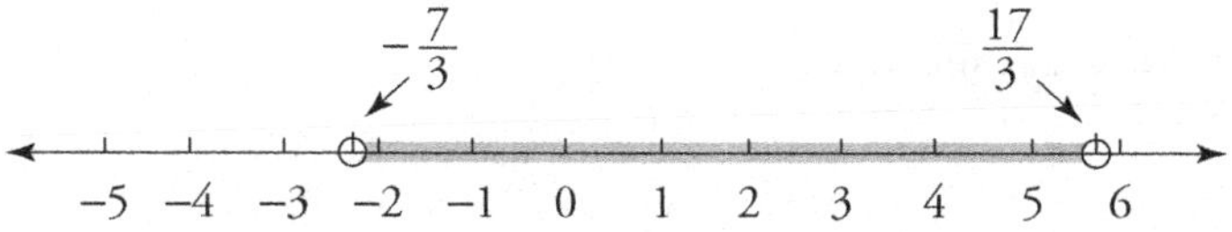

Example 2: Solve this disjunction for x: $|5x + 3| > 2$.

$$\begin{aligned} 5x+3 &< -2 & \text{or} \quad 5x+3 &> 2 \\ 5x &< -5 & 5x &> -1 \\ x &< -1 & x &> -\frac{1}{5} \end{aligned}$$

The solution set is $\left\{x \mid x < -1 \text{ or } x > -\frac{1}{5}\right\}$. The graph of the solution set is shown in the following graph: x is less than -1 or greater than $-\frac{1}{5}$.

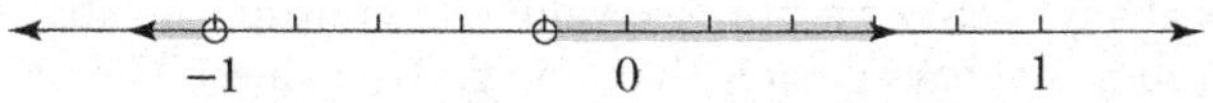

The interval notation is $(-\infty, -1) \cup \left(-\frac{1}{5}, \infty\right)$.

Example 3: Solve for x: $|2x + 11| < 0$.

There is no solution for this inequality because the absolute value of a number is never negative.

Example 4: Solve for x: $|2x + 11| > 0$.

The solution is all real numbers *except* for the solution to $2x + 11 = 0$. Therefore,

$$2x + 11 \neq 0$$
$$2x \neq -11$$
$$x \neq -\frac{11}{2}$$

The solution of the set is $\left\{x \mid x \text{ is a real number, } x \neq -\frac{11}{2}\right\}$. The graph of the solution set is shown here.

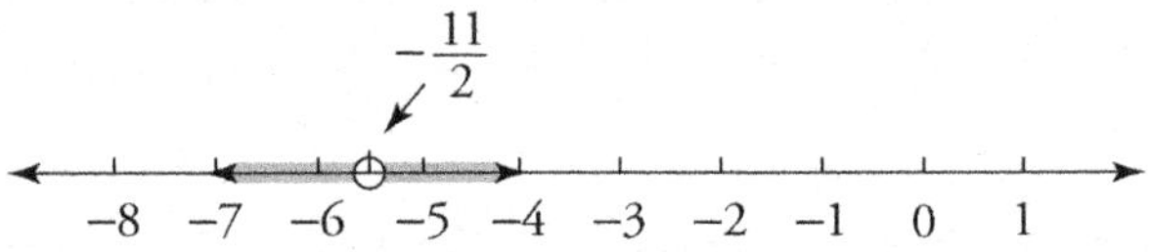

Example 5: Solve for x: $7|3x + 2| + 5 > 4$.

First, isolate the expression involving the absolute value symbol.

$$7|3x + 2| + 5 > 4$$
$$7|3x + 2| > -1$$
$$|3x + 2| > -\frac{1}{7}$$

The solution set is all real numbers. (***Note:*** The absolute value of any number is always zero or a positive value. Therefore, the absolute value of any number is always greater than a negative value.) The graph of the solution set is shown here.

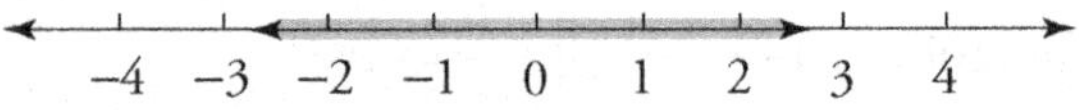

Chapter Check-Out

Questions

1. Solve for x: $5(3x - 7) = 6(5x + 9)$.
2. Final velocity (v_f) = initial velocity (v_i) + acceleration (a) × time (t). Solve the formula for a:

$$v_f = v_i + at$$

3. Wendy has no more than 68 feet of fencing to enclose her rectangular vegetable garden. The length of the garden is 4 feet longer than the width. What are the largest dimensions that her garden can be?
4. Solve for y: $3y + 2 < 17$ and $2y - 3 > -15$.
5. Solve for x: $|5x - 3| < 14$.
6. Find the smallest three consecutive integers whose sum is greater than 27.
7. Dina can spend no more than $500 to host a staff luncheon. She will be charged $60 to rent a room, plus $15 per person. What is the maximum number of people she can host at the luncheon?

Answers

1. $x = -\dfrac{89}{15}$
2. $a = \dfrac{v_f - v_i}{t}$
3. The width must be less than or equal to 15. The length must be less than or equal to 19.
4. $\{y | -6 < y < 5\}$
5. $\left\{x \middle| -\dfrac{11}{5} < x < \dfrac{17}{5}\right\}$
6. The smallest integers are 9, 10, and 11.
7. Dina can host a maximum of 29 people.

Chapter 2

SEGMENTS, LINES, AND INEQUALITIES IN THE COORDINATE PLANE

Chapter Check-In

- ❑ Understanding the rectangular coordinate system
- ❑ Deriving and applying the distance formula
- ❑ Understanding and applying the midpoint formula
- ❑ Understanding slopes and intercepts
- ❑ Using slopes of parallel and perpendicular lines
- ❑ Graphing linear equations and determining equations of lines
- ❑ Graphing linear inequalities

Common Core Standard: Seeing the Structure in Expressions

Understand linear equations and inequalities in two variables, understand slopes, and graph linear equations and inequalities. Represent and solve equations and inequalities graphically (A.REI.11).

Chapter 1 presented linear equations and inequalities containing one variable. This chapter introduces linear equations and inequalities containing two variables. Two-variable equations are graphed on two number lines that intersect at right angles to form the **coordinate plane,** also referred to as the ***x*-axis** (horizontal number line) and the ***y*-axis** (vertical number line). Together these axes are called *coordinate axes.*

The location of every point on a graph can be determined by two coordinates, written as an ordered pair, (x, y). These are also known as *Cartesian coordinates,* after the French mathematician René Descartes, who is credited with their invention. If the slope and y-intercept, or the coordinates of two points on a linear graph are known, the equation of the line can be determined. The concepts of slope and intercepts are introduced in this chapter.

Rectangular Coordinate System

Following are coordinate system terms with which you should be familiar:

- **Coordinates of a point.** Each point on a *number line* is assigned a number. In the same way, each point in a *plane* is assigned a pair of numbers called the coordinates of the point.
- ***x*-axis; *y*-axis.** To locate points in a plane, two perpendicular lines are used: a horizontal line called the x-axis and a vertical line called the y-axis.
- **Coordinate plane.** The x-axis, the y-axis, and all the points in their plane are referred to as a coordinate plane.
- **Ordered pairs.** Every point in a coordinate plane is named by a pair of numbers whose order is important. This pair of numbers, written in parentheses and separated by a comma, is the ordered pair for the point. The x-coordinate is written first, followed by the y-coordinate.
- **Origin.** The point of intersection of the x-axis and the y-axis is called the origin. The ordered pair for the origin is (0, 0).
- ***x*-coordinate.** The number to the left of the comma in an ordered pair is the x-coordinate of the point and indicates the amount of horizontal movement from the origin. The movement is to the right if the number is positive and to the left if the number is negative. The x-coordinate is also known as the *abscissa.*
- ***y*-coordinate.** The number to the right of the comma in an ordered pair is the y-coordinate of the point and indicates the amount of vertical movement from the origin. The movement is up if the number is positive and down if the number is negative. The y-coordinate is also known as the *ordinate.*
- **Quadrants.** The x-axis and y-axis separate the coordinate plane into four regions called quadrants. The upper right quadrant is quadrant I, the upper left quadrant is quadrant II, the lower left quadrant is quadrant III, and the lower right quadrant is quadrant IV. Notice that, as shown in the figure that follows,

  in quadrant I, x is always positive and y is always positive $(+, +)$

  in quadrant II, x is always negative and y is always positive $(-, +)$

in quadrant III, x is always negative and y is always negative $(-, -)$
in quadrant IV, x is always positive and y is always negative $(+, -)$

The point associated with an ordered pair of real numbers is called the graph of the ordered pair, or the plot of that point.

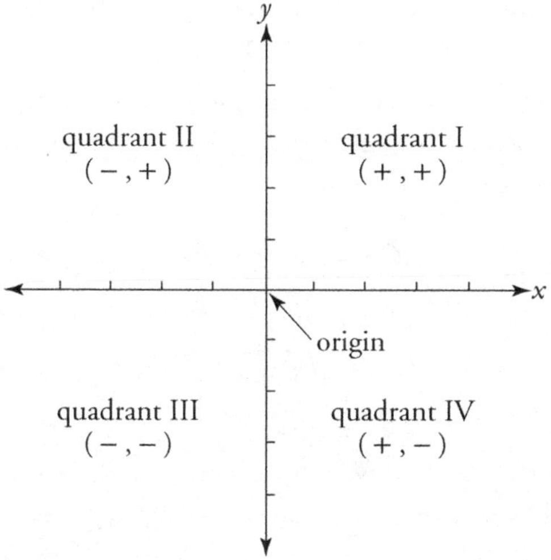

Distance Formula

In the following graph of a right triangle, A has coordinates (2, 2), B has coordinates (5, 2), and C has coordinates (5, 6).

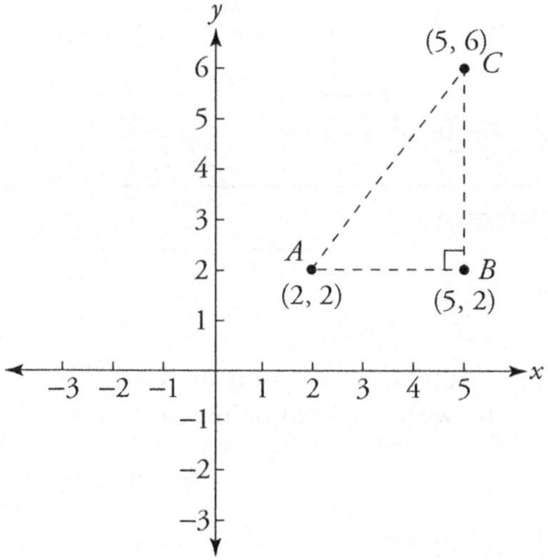

To find the length *AB* or *BC*, only simple subtraction is necessary. The length of line segment *AB* is found by subtracting the *x*-coordinates of *A* and *B*. That is, 5 − 2 = 3 for the *x* distance. The length of line segment *BC* is found by subtracting the *y*-coordinates of *B* and *C*. That is, 6 − 2 = 4 for the *y* distance.

Note that the notation "*AB*" is used to represent the numerical length of the line segment from point *A* to point *B*.

To find the length *AC*, however, simple subtraction is not sufficient. Triangle *ABC* is a right triangle with $\overline{AC}$ being the hypotenuse. Therefore, by the Pythagorean theorem,

$$\begin{aligned} AC^2 &= AB^2 + BC^2 \\ AC &= \sqrt{AB^2 + BC^2} \\ &= \sqrt{3^2 + 4^2} \\ &= \sqrt{9+16} \\ &= \sqrt{25} \\ &= 5 \end{aligned}$$

From the Pythagorean theorem, we derive the distance formula, which is nothing more than a different format for the Pythagorean theorem. If *A* is represented by the ordered pair (x_1, y_1) and *C* is represented by the ordered pair (x_2, y_2), then

$$AB = |x_2 - x_1| \qquad \text{and} \qquad BC = |y_2 - y_1|$$

Then $AC = \sqrt{(x_2 - x_1)^2 + (y_2 - y_1)^2}$.

Distance Formula

$$d = \sqrt{(x_2 - x_1)^2 + (y_2 - y_1)^2}$$

In the preceding formula, *d* stands for distance. This formula is used to find the distance between any two points in the coordinate plane.

Example 1: Use the distance formula to find the distance between the points with coordinates (−3, 4) and (5, 2).

Substitute (−3, 4) for (x_1, y_1) and (5, 2) for (x_2, y_2). Then

$$\begin{aligned} d &= \sqrt{[5-(-3)]^2 + (2-4)^2} \\ &= \sqrt{8^2 + (-2)^2} \\ &= \sqrt{64+4} \\ &= \sqrt{68} \\ &= \sqrt{4}\sqrt{17} \\ &= 2\sqrt{17} \end{aligned}$$

Midpoint Formula

A line segment has a finite length; therefore, its midpoint can always be determined. Numerically, the midpoint of a segment can be considered to be the average of its endpoints. This concept should help in remembering a formula for finding the midpoint of a segment given the coordinates of its endpoints. Recall that the average of two numbers is found by dividing their sum by 2.

Midpoint Formula

$$M = \left(\frac{x_1 + x_2}{2}, \frac{y_1 + y_2}{2}\right)$$

Example 1: R is the midpoint between $Q(-9, -1)$ and $T(-3, 7)$. Find its coordinates.

By the midpoint formula,

$$\begin{aligned} M &= \left(\frac{x_1 + x_2}{2}, \frac{y_1 + y_2}{2}\right) \\ R &= \left(\frac{-9+(-3)}{2}, \frac{-1+7}{2}\right) \\ &= \left(\frac{-12}{2}, \frac{6}{2}\right) \\ &= (-6, 3) \end{aligned}$$

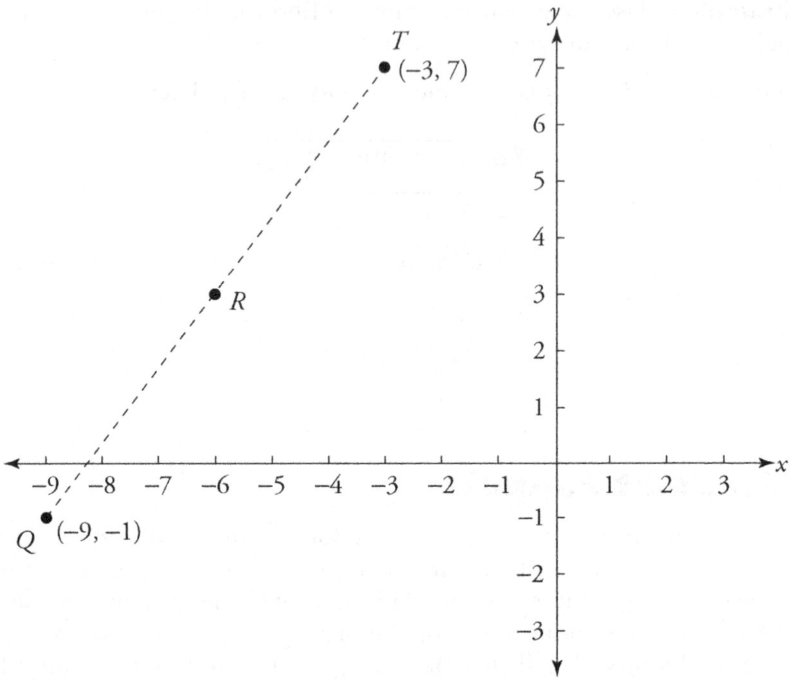

Slope of a Line

The **slope of a line** is a measurement of the steepness and direction of a nonvertical line. When a line slants from lower left to upper right, the slope is a positive number. Item (a) in the following figure shows a line with a positive slope. When a line slants from upper left to lower right, the slope is a negative number (b). The *x*-axis or any line parallel to the *x*-axis has a slope of zero; that is, a horizontal line has a slope of zero (c). The *y*-axis or any line parallel to the *y*-axis has no defined slope; that is, a vertical line has an undefined slope (d), or no slope.

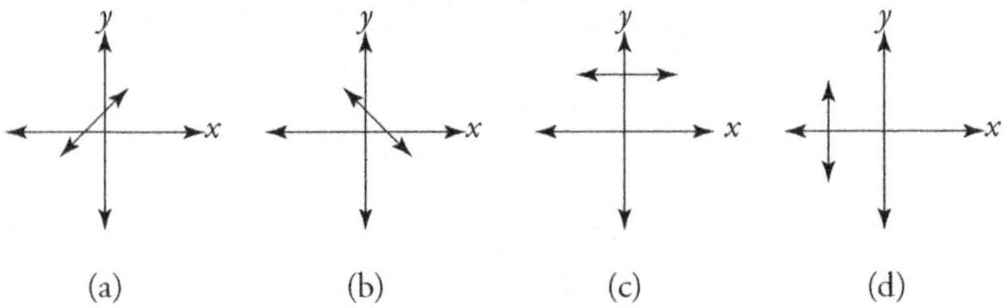

If m represents the slope of a line and A and B are points lying on that line with coordinates (x_1, y_1) and (x_2, y_2), respectively, then the slope of the line passing through A and B is given by the following formula.

Slope Formula

$$m = \frac{y_2 - y_1}{x_2 - x_1},\ x_2 \neq x_1$$

Since A and B cannot be points on a vertical line, x_1 and x_2 cannot be equal to each other. If $x_1 = x_2$, then the line is vertical and the slope is undefined.

A slope is also referred to as being the rate of change between two values. It is the ratio of the change in the y-variable to the change in the x-variable.

Example 1: In the following figure, find the slopes of the lines a, b, c, and d.

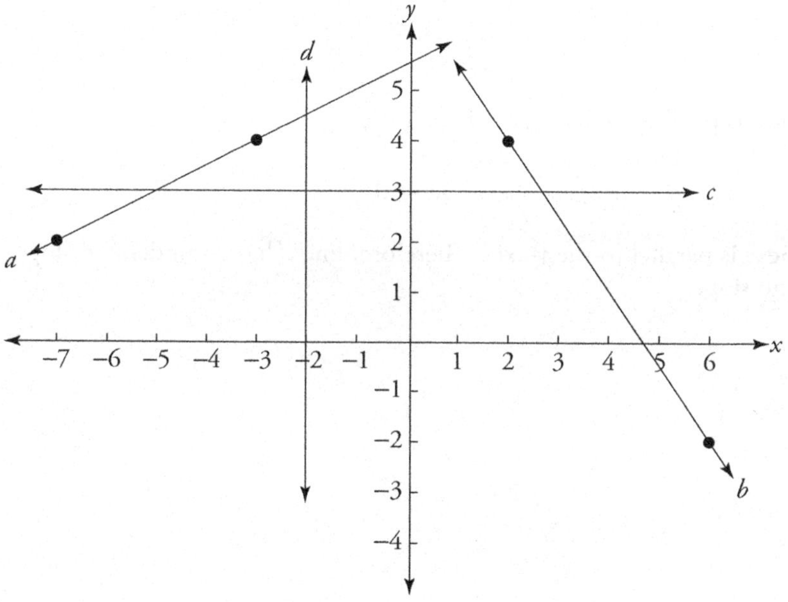

Line *a* passes through the points (−7, 2) and (−3, 4).

$$\begin{aligned} m &= \frac{y_2 - y_1}{x_2 - x_1} \\ &= \frac{4-2}{-3-(-7)} \\ &= \frac{2}{4} \\ &= \frac{1}{2} \end{aligned}$$

Line *b* passes through the points (2, 4) and (6, −2).

$$\begin{aligned} m &= \frac{y_2 - y_1}{x_2 - x_1} \\ &= \frac{-2-4}{6-2} \\ &= \frac{-6}{4} \\ &= -\frac{3}{2} \end{aligned}$$

Line *c* is parallel to the *x*-axis. Therefore,

$$m = 0$$

Line *d* is parallel to the *y*-axis. Therefore, line *d* has an undefined slope, or no slope.

Example 2: A line passes through $(-5, 8)$ with a slope of $\frac{2}{3}$. If another point on this line has coordinates $(x, 12)$, find x.

$$m = \frac{y_2 - y_1}{x_2 - x_1}$$

$$\frac{2}{3} = \frac{12 - 8}{x - (-5)}$$

$$\frac{2}{3} = \frac{4}{x + 5}$$

$$2(x + 5) = 4(3)$$

$$2x + 10 = 12$$

$$x = 1$$

Slopes of Parallel and Perpendicular Lines

Parallel lines have equal slopes. Conversely, if two different lines have equal slopes, they are parallel. If two nonvertical lines are perpendicular, then their slopes are negative reciprocals (opposite reciprocals) of one another, or the product of their slopes is -1. Conversely, if the slopes of two lines are opposite reciprocals of each other, or the product of their slopes is -1, then the lines are nonvertical perpendicular lines. Because horizontal and vertical lines are always perpendicular, then lines having a zero slope and an undefined slope are perpendicular.

Example 1: Fill in the blanks.

If line l has slope $\frac{3}{4}$, then any line parallel to line l will have slope **(a)** _____, and any line perpendicular to line l will have slope **(b)** _____.

(a) $\frac{3}{4}$

(b) $-\frac{4}{3}$

Graphing Linear Equations and Writing Equations of Lines

Equations involving one or two variables can be graphed on any *xy*-coordinate plane. In general, it is true that

- if a point lies on the graph of an equation, then its coordinates make the equation a true statement, and
- if the coordinates of a point make an equation a true statement, then the point lies on the graph of the equation.

The graphs of linear equations are always lines. All linear equations can be written in the form $Ax + By = C$, where A, B, and C are real numbers, and A and B are not both zero. Furthermore, to be in standard form, A has to be a positive number. Following are examples of linear equations and their respective A, B, and C values.

Note that the equations $x = -6$ and $y = 7$ appear to contain only one variable. These equations are two special cases of lines in a coordinate plane. You can also write them as follows: $1x + 0y = -6$ and $0x + 1y = 7$, respectively. When $B = 0$, the line will be vertical, and when $A = 0$, the line will be horizontal.

Note that nonvertical lines are also referred to as *linear functions.*

$x + y = 0$	$3x - 4y = 9$	$x = -6$	$y = 7$
$A = 1$	$A = 3$	$A = 1$	$A = 0$
$B = 1$	$B = -4$	$B = 0$	$B = 1$
$C = 0$	$C = 9$	$C = -6$	$C = 7$

Following are terms with which you should be familiar:

- **Standard form (of a line).** The form $Ax + By = C$ for the equation of a line is known as the standard form for the equation of a line.
- **Slope-intercept form.** The form $y = mx + b$ for the equation of a line is known as the slope-intercept form for the equation of the line whose slope is m and whose y-intercept is b.
- **Function form.** The form $f(x) = mx + b$ for the equation of a line is known as function form, where the y value of an equation in slope-intercept form is expressed as a function of the corresponding x value. ***Note:*** Only nonvertical lines are functions.

- **Point-slope form.** The form $y - y_1 = m(x - x_1)$ for the equation of a line is known as point-slope form, where m is the slope of the line and (x_1, y_1) is a point on that line. ***Note:*** This form can only be used for nonvertical lines, as its derivation comes from the definition of slope; see Examples 1–4 in this section.
- ***x*-intercept.** The *x*-intercept of a graph is the point at which the graph will intersect the *x*-axis. It always has a *y*-coordinate of zero. A horizontal line that is not the *x*-axis has no *x*-intercept.
- ***y*-intercept.** The *y*-intercept of a graph is the point at which the graph will intersect the *y*-axis. It always has an *x*-coordinate of zero. A vertical line that is not the *y*-axis has no *y*-intercept.

One of the many methods to graph a linear equation is to find solutions by giving a value to one variable and solving the resulting equation for the other variable. A minimum of two points is necessary to graph a linear equation. This method allows you to choose your own values.

Example 1: Draw the graph of $2x + 3y = 12$ by finding two random points.

To do this, select a value for one variable, then substitute this into the equation and solve for the other variable. Do this a second time with new values to get a second point.

Let $x = 2$, then find y.

$$\begin{aligned} 2x+3y&=12 \\ 2(2)+3y&=12 \\ 4+3y&=12 \\ 3y&=8 \\ y&=\frac{8}{3} \end{aligned}$$

Therefore, the ordered pair $\left(2, \frac{8}{3}\right)$ belongs on the graph.

Let $y = 6$, then find x.

$$\begin{aligned} 2x+3y&=12 \\ 2x+3(6)&=12 \\ 2x+18&=12 \\ 2x&=-6 \\ x&=-3 \end{aligned}$$

Therefore, the ordered pair $(-3, 6)$ belongs on the graph.

Graph these points and then connect them to graph the line of $2x + 3y = 12$.

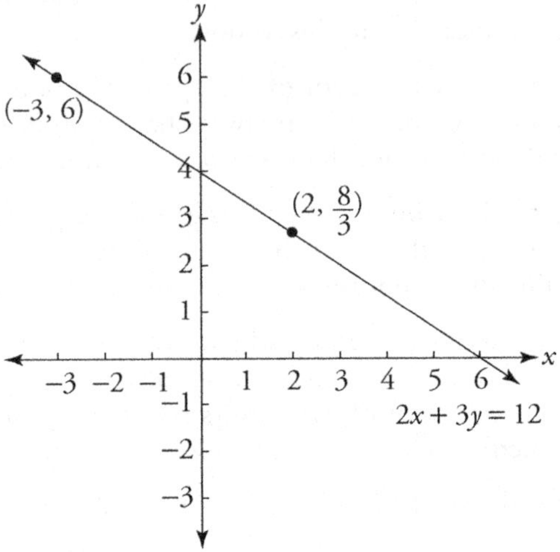

Another method used to graph a line is by locating its intercepts, as shown in the following example.

Example 2: Draw the graph of $2x + 3y = 12$ by finding the x-intercept and the y-intercept.

The x-intercept has a y-coordinate of zero. Substituting zero for y, the resulting equation is

$$2x + 3(0) = 12$$

Now, solving for x,

$$2x = 12$$

$$x = 6$$

The x-intercept is at $(6, 0)$.

The y-intercept has an x-coordinate of zero. Substituting zero for x, the resulting equation is

$$2(0) + 3y = 12$$

Now, solving for y,

$$3y = 12$$

$$y = 4$$

The y-intercept is at (0, 4).

The line now can be graphed by plotting these two points and drawing the line they determine.

Notice that graphs for Examples 1 and 2 are exactly the same. Both are the graph of the line $2x + 3y = 12$.

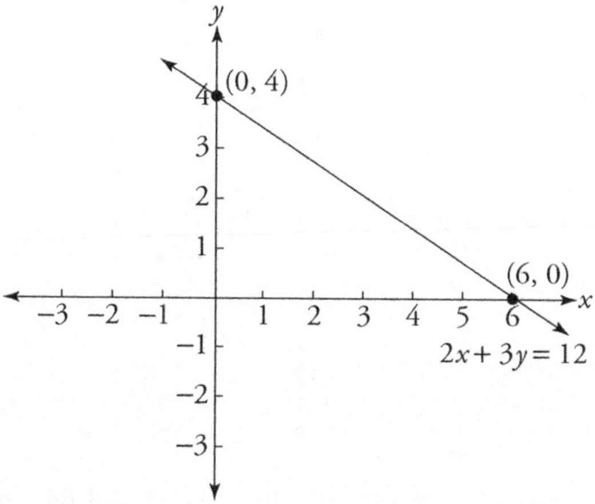

Graphing linear equations containing only one variable requires the understanding of "Graphing Linear Equations and Writing Equations of Lines," as detailed on pp. 36–37. Two such examples are shown below in Examples 3 and 4.

Example 3: Draw the graph of $x = 2$.

As shown in the following figure, $x = 2$ is a vertical line whose x-coordinate is always 2.

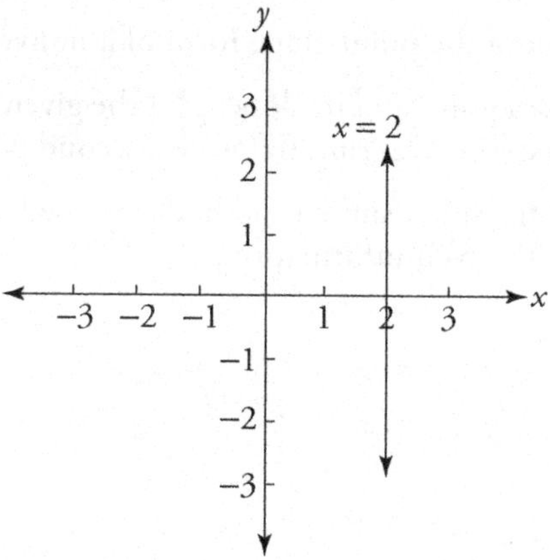

Example 4: Draw the graph of $y = -1$.

As shown in the following figure, $y = -1$ is a horizontal line whose y-coordinate is always -1.

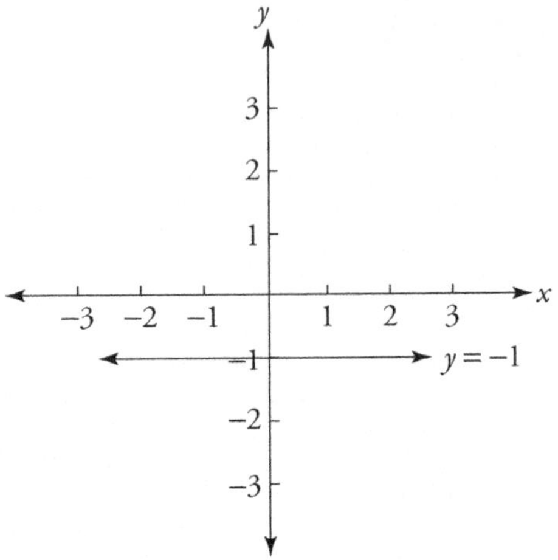

Suppose that A is a particular point called (x_1, y_1) and B is any point called (x, y). Then the slope of the line through A and B is defined by

$$\frac{y - y_1}{x - x_1} = m$$

Apply the cross products rule, and the equation becomes

$$y - y_1 = m(x - x_1)$$

This is the derivation of the point-slope form of a nonvertical line.

To graph a line given a point and its slope, plot the given point and apply the definition of slope (rise over run) to locate a second point on that line.

In addition to graphing lines, similar methods are used to determine the equation of a line with given information.

Example 5: Find the equation of the line containing the points (−3, 4) and (7, 2) and write the equation in both point-slope form and standard form.

For the point-slope form, first find the slope, m.

$$\begin{aligned} m &= \frac{y_2 - y_1}{x_2 - x_1} \\ &= \frac{2-4}{7-(-3)} \\ &= -\frac{2}{10} \\ &= -\frac{1}{5} \end{aligned}$$

Now, choose either given point, say (−3, 4), and substitute the x and y values into the point-slope form.

$$y - y_1 = m(x - x_1)$$

$$y - 4 = -\frac{1}{5}[x - (-3)] \text{ or } y - 4 = -\frac{1}{5}(x+3)$$

For the standard form, begin with the point-slope form and clear it of fractions by multiplying both sides by the least common denominator.

$$y - 4 = -\frac{1}{5}(x+3)$$

Multiply both sides by 5.

$$\begin{aligned} 5(y-4) &= 5\left[-\frac{1}{5}(x+3)\right] \\ 5y - 20 &= -(x+3) \\ 5y - 20 &= -x - 3 \end{aligned}$$

Get x and y on one side and the constants on the other side by adding x to both sides and adding 20 to both sides. Make sure that A is a positive number.

$$x + 5y = 17$$

Nonvertical lines

A nonvertical line written in standard form is $Ax + By = C$, with $B \neq 0$. Solve this equation for y.

$$\begin{aligned} Ax + By &= C \\ By &= -Ax + C \\ y &= -\frac{A}{B}x + \frac{C}{B} \end{aligned}$$

The value $-\frac{A}{B}$ becomes the slope of the line, and $\frac{C}{B}$ becomes the y-intercept value. When $-\frac{A}{B}$ is replaced with m and $\frac{C}{B}$ is replaced with b, the equation becomes $y = mx + b$. This is the slope-intercept form of a nonvertical line.

Example 6: Find the slope and y-intercept value of the line with the equation $3x - 4y = 20$.

Solve for y.

$$\begin{aligned} 3x - 4y &= 20 \\ -4y &= -3x + 20 \\ y &= \frac{3}{4}x - 5 \end{aligned}$$

Therefore, the slope of the line is $\frac{3}{4}$ and the y-intercept value is -5.

Example 7: Draw the graph of the equation $y = -\frac{3}{4}x + 5$.

The equation is in slope-intercept form. The slope is $-\frac{3}{4}$ and the y-intercept is (0, 5). From this, the graph can be quickly drawn. Because the slope is negative, the line is slanting to the upper left/lower right. Begin with the y-intercept (0, 5) and use the slope to find additional points. Either go up 3 and left 4 or go down 3 and right 4. Now, label these points and connect them, as shown in the following graph.

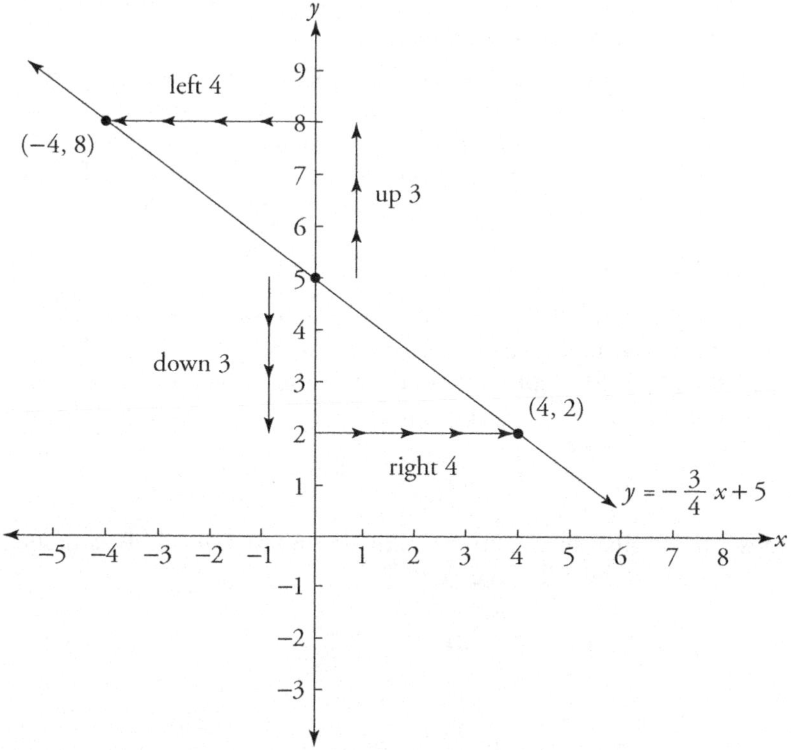

Graphs of Linear Inequalities

A **linear inequality** in standard form is a mathematical sentence in one of the following forms:

$Ax + By < C$

$Ax + By > C$

$Ax + By \leq C$

$Ax + By \geq C$

To graph such sentences, follow these steps:

1. **Graph the linear equation $Ax + By = C$.** This line becomes a boundary line for the graph. If the original inequality is $<$ or $>$, the boundary line is drawn as a dashed line, since the points on the line do not make the original sentence true. If the original inequality is $\leq$ or $\geq$, the boundary line is drawn as a solid line, since the points on the line will make the original inequality true.

2. **Select a point not on the boundary line and substitute its *x* and *y* values into the original inequality.**
3. **Shade the appropriate area.** If the resulting sentence is true, then shade the region where that test point is located, indicating that all the points on that side of the boundary line will make the original sentence true. If the resulting sentence is false, then shade the region on the side of the boundary line opposite that where the test point is located.

Example 1: Graph $3x + 4y < 12$.

First, draw the graph of $3x + 4y = 12$. If you use the *x*-intercept and *y*-intercept method, you get *x*-intercept (4, 0) and *y*-intercept (0, 3). If you use the slope-intercept method, the equation, when written in slope-intercept form ($y = mx + b$), becomes

$$y = -\frac{3}{4}x + 3$$

Because the original inequality is <, the boundary line will be a dashed line, as shown in the following graph.

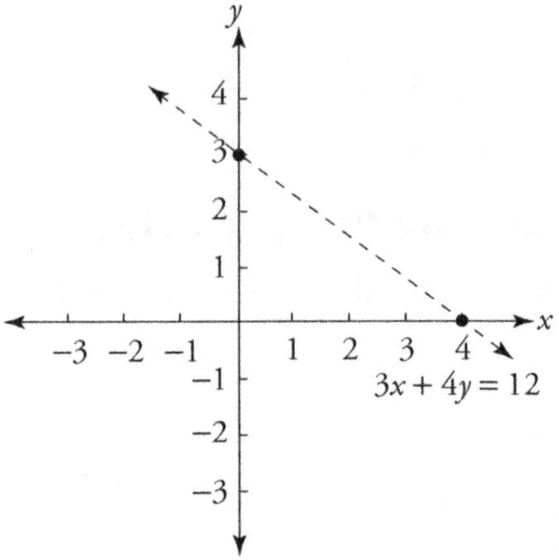

Now select a point not on the boundary, say (0, 0). Substitute this into the original inequality:

$$3x + 4y < 12$$

$$3(0) + 4(0) \overset{?}{<} 12$$

$$0 + 0 \overset{?}{<} 12$$

$$0 < 12 \checkmark$$

This is a true statement. This means that the "(0, 0) side" of the boundary line is the desired region to be shaded. Now, shade that region as shown in the following graph. The shading is below the line.

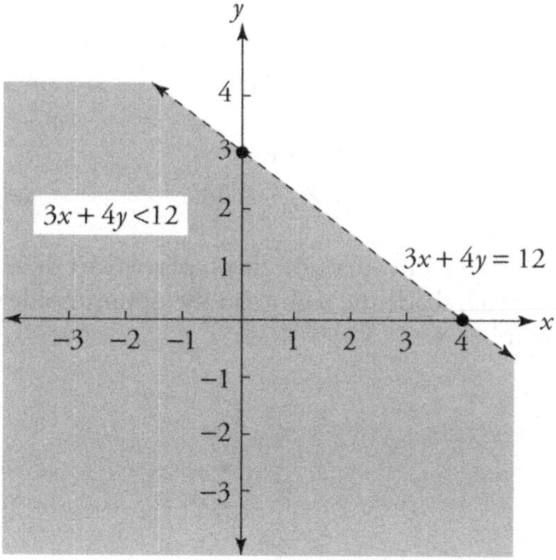

Example 2: Graph $y \geq 2x + 3$.

First, graph $y = 2x + 3$, as shown below.

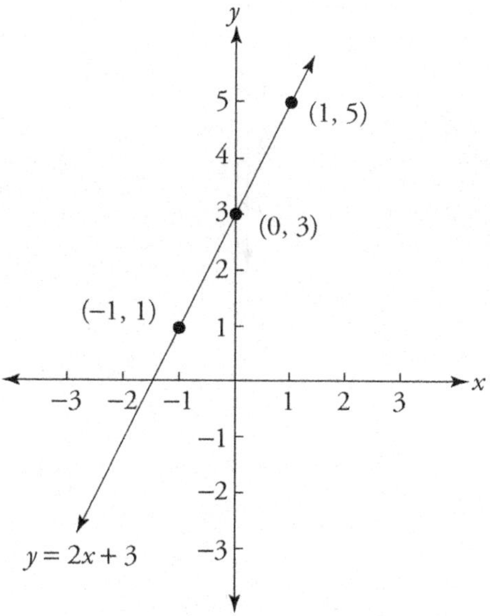

Notice that the boundary is a solid line because the original inequality is $\geq$. Now, select a point not on the boundary, say (2, 1), and substitute its x and y values into $y \geq 2x + 3$.

$$y \geq 2x+3$$

$$1 \overset{?}{\geq} 2(2)+3$$

$$1 \overset{?}{\geq} 4+3$$

$$1 \ngeq 7$$

This is not a true statement. Because this replacement does not make the original sentence true, shade the region on the opposite side of the boundary line, as shown below.

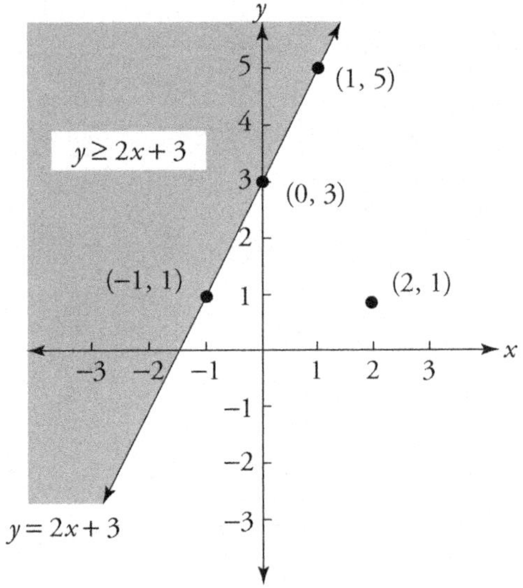

Example 3: Graph $x < 2$.

The graph of $x = 2$ is a vertical line whose points all have the x-coordinate of 2.

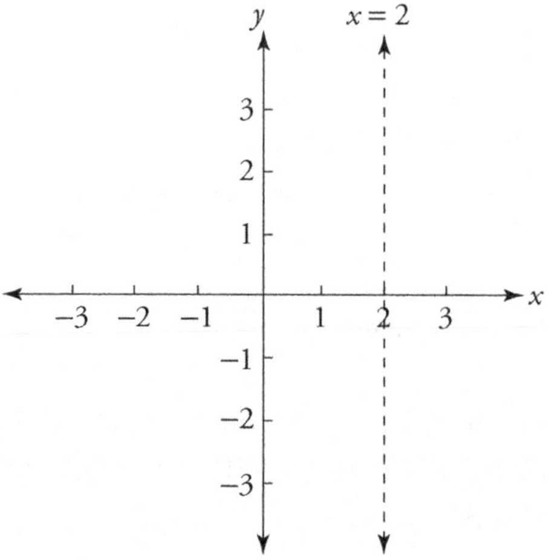

Select a point not on the boundary, say (0, 0). Substitute the x value into $x < 2$.

$$x < 2$$
$$0 < 2 \checkmark$$

This is a true statement. Therefore, shade in the "(0, 0) side" of the boundary line, as shown below.

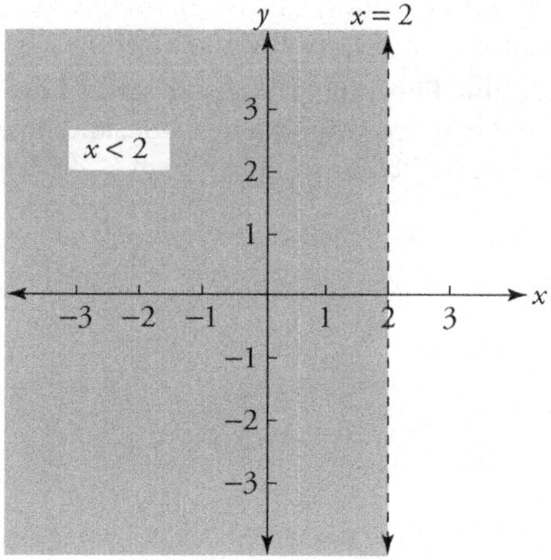

Chapter Check-Out

Questions

1. A line segment connects (1, 3) and (6, 7). Using the distance formula, find the length of that segment.
2. A line segment connects A $(-5, -9)$ with B (7, 21). Point R is at the midpoint of that line. What are R's coordinates?
3. A line passes through $(-5, 12)$ and (4, 7). What is its slope?
4. What are the coordinates of the x- and y-intercepts in the following equation?

$$3x + 2y = 24$$

Questions 5–7 refer to the following information about parallel and perpendicular lines.

Find the slope of each of the following lines to classify each set of equations as being lines that are parallel, perpendicular, or neither.

5. $-5x - y = 3$ and $10x + 2y = 8$
6. $x - 3y = 6$ and $6x + 2y = 3$
7. $10x - 5 = 3y$ and $2x + 3y = 5$
8. A swimming pool is draining steadily at 3 gallons of water per hour since midnight. There were 1,000 gallons of water at 5:00 a.m. Let x represent the number of hours since midnight, and let y represent the number of gallons in the swimming pool. Write the equation that shows the amount of water remaining in the swimming pool at time x.
9. Does the point (4, 2) lie in the region defined by $2x - 3y \leq 6$? Show your response both graphically and algebraically.
10. When graphing linear inequalities, describe how to determine which region of the graphed equation to shade, and how to determine if the line should be dashed or solid.

Answers

1. $\sqrt{41}$
2. (1, 6)
3. $-\dfrac{5}{9}$
4. (8, 0) and (0, 12)
5. 1st equation: $m = -5$; 2nd equation: $m = -5$. The lines are parallel.
6. 1st equation: $m = \dfrac{1}{3}$; 2nd equation: $m = -3$. The lines are perpendicular.
7. 1st equation: $m = \dfrac{10}{3}$; 2nd equation: $m = -\dfrac{2}{3}$. The lines are neither parallel nor perpendicular.
8. $y = -3x + 1{,}015$
9. Yes

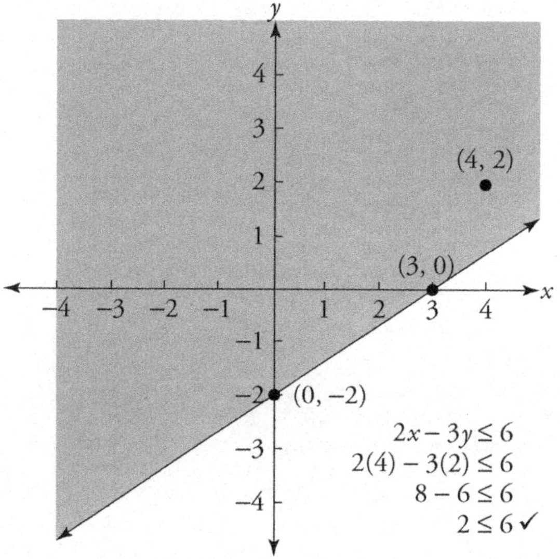

10. One way is to choose a point randomly and substitute the x and y values into the inequality. The line will be solid if the inequality symbol is $\leq$ or $\geq$. The line will be dashed if the inequality symbol is $<$ or $>$.

Chapter 3

LINEAR SYSTEMS IN TWO VARIABLES

Chapter Check-In

- ❑ Solving systems of equations by graphing
- ❑ Solving systems of equations by substitution
- ❑ Solving systems of equations by elimination
- ❑ Solving systems of equations using matrices
- ❑ Solving systems of equations using determinants (Cramer's Rule)
- ❑ Solving word problems that translate into linear systems
- ❑ Solving systems of linear inequalities by graphing
- ❑ Understanding and using linear programming

Common Core Standard: Seeing the Structure in Expressions

Solve systems of linear equations exactly and approximately (e.g., with graphs), focusing on pairs of linear equations in two variables (A.REI.6). Solve a simple system consisting of a linear equation and a quadratic equation in two variables algebraically and graphically (A.REI.7).

Earlier chapters covered solving one equation with a single variable. This chapter looks at ways to solve systems of equations and inequalities that contain two variables. A **system** is a set of equations or inequalities sometimes containing as many variables as there are sentences. They are called systems because there will be more than one equation containing the same variables. Solving a system means finding the values of each variable that simultaneously satisfy each equation or inequality in the system. The ways to solve systems include graphing, substitution, elimination, matrices, and determinants. When graphing to solve the system, the solution is the point or the region that simultaneously satisfies the sentences in the system. Certain systems are more easily solved using one method instead of another.

Solving Systems of Equations by Graphing

Example 1: Solve this system of equations by graphing.

$$\begin{cases} 4x+3y=6 \\ 2x-5y=16 \end{cases}$$

To solve using graphing, graph both equations on the same set of coordinate axes and see where the graphs intersect. An equation is satisfied only by the ordered pairs along the corresponding line, and nowhere else. The ordered pair at the point of intersection becomes the solution. At the intersection, both equations are true.

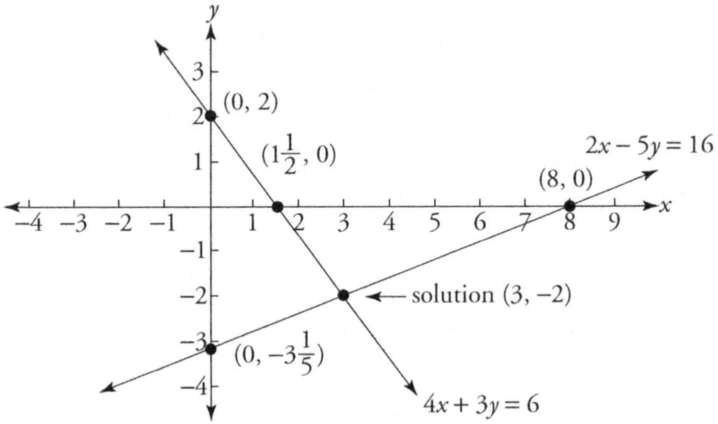

Check the solution.

$4x+3y=6$	$2x-5y=16$
$4(3)+3(-2)\stackrel{?}{=}6$	$2(3)-5(-2)\stackrel{?}{=}16$
$12-6\stackrel{?}{=}6$	$6+10\stackrel{?}{=}16$
$6=6$ ✓	$16=16$ ✓

The solution is $x = 3, y = -2$; this is also written as the point $(3, -2)$.

Solving systems of equations by graphing is limited to equations in which the solution lies close to the origin and consists of integers. Otherwise, it would be hard to both graph the lines and see what the ordered pair is at the intersection. For those reasons, graphing is used least frequently of all the solution methods. Here are two things to keep in mind:

- **Dependent system.** If the two graphs coincide—that is, if they are two versions of the same equation—then the system is called a dependent system, and its solution can be expressed as all values of the variables that satisfy either of the two original equations.
- **Inconsistent system.** If the two graphs are parallel—that is, if there is no point of intersection—then the system is called an inconsistent system, and its solution is expressed as an empty set { }, or the null set, ∅.

Solving Systems of Equations by Substitution

To solve systems using substitution, follow these steps:

1. Select one equation and solve it for one of its variables.
2. In the other equation, substitute for the variable just solved.
3. Solve the new equation.
4. Substitute the value found into any equation involving both variables and solve for the other variable.
5. Check the solution in both original equations.

Usually, when using the substitution method, one equation and one of the variables leads to a quick solution more readily than the other. That's illustrated by the selection of x and the second equation in the following example.

Example 1: Solve this system of equations by using substitution.

$$\begin{cases} 3x+4y=-5 \\ 2x-3y=8 \end{cases}$$

Solve for x in the second equation.

$$\begin{aligned} 2x-3y&=8 \\ 2x&=3y+8 \\ x&=\frac{3}{2}y+4 \end{aligned}$$

Substitute $\frac{3}{2}y+4$ for x in the other equation.

$$3x+4y=-5$$

$$3\left(\frac{3}{2}y+4\right)+4y=-5$$

Solve this new equation.

$$\frac{9}{2}y+12+4y=-5$$

$$\frac{17}{2}y=-17$$

$$y=-2$$

Substitute the value found for y into any equation involving both variables.

$$x=\frac{3}{2}y+4$$

$$x=\frac{3}{2}(-2)+4$$

$$x=-3+4$$

$$x=1$$

Check the solution in both original equations.

$$3x+4y=-5 \qquad\qquad 2x-3y=8$$

$$3(1)+4(-2)\stackrel{?}{=}-5 \qquad\qquad 2(1)-3(-2)\stackrel{?}{=}8$$

$$3-8\stackrel{?}{=}-5 \qquad\qquad 2+6\stackrel{?}{=}8$$

$$-5=-5\ \checkmark \qquad\qquad 8=8\ \checkmark$$

The solution is $x = 1, y = -2$; this is also written as the point $(1, -2)$.

If the substitution method produces a sentence that is always true, such as $0 = 0$ or $7 = 7$, then the system is dependent, and values that satisfy either original equation is a solution. If the substitution method produces a sentence that is always false, such as $0 = 5$ or $7 = -9$, then the system is inconsistent, and there is no solution, or $\varnothing$.

Solving Systems of Equations by Elimination

To solve systems using elimination, follow these steps:

1. Arrange both equations in standard form (see p. 36), placing like variables and constants one above the other.
2. Choose a variable to eliminate, and with a proper choice of multiplication, arrange so that the coefficients of that variable are opposites of one another.
3. Add the equations, leaving one equation with one variable.
4. Solve for the remaining variable.
5. Substitute the value found in Step 4 into any original equation and solve for the other variable.
6. Check the solution in both original equations.

Example 1: Solve this system of equations by using elimination.

$$\begin{cases} 2x-3=-5y \\ \quad -2y=-3x+1 \end{cases}$$

Arrange both equations in standard form, placing like terms one above the other.

$$\begin{aligned} 2x+5y&=3 \\ 3x-2y&=1 \end{aligned}$$

Select a variable to eliminate, say y.

The coefficients of y are 5 and -2. These both divide into 10. Arrange so that the coefficient of y is 10 in one equation and -10 in the other. To do this, multiply the top equation by 2 and the bottom equation by 5.

$$2x+5y=3 \xrightarrow{\text{multiply}(2)} 4x+10y=6$$

$$3x-2y=1 \xrightarrow{\text{multiply}(5)} 15x-10y=5$$

Add the new equations, eliminating *y*.

$$\begin{array}{r} 4x+10y=6 \\ 15x-10y=5 \\ \hline 19x \qquad =11 \end{array}$$

Solve for the remaining variable.

$$x=\frac{11}{19}$$

Substitute for *x* and solve for *y*.

$$\begin{aligned} 2x+5y&=3 \\ 2\left(\frac{11}{19}\right)+5y&=3 \\ \frac{22}{19}+5y&=3 \\ 5y&=\frac{35}{19} \\ y&=\frac{7}{19} \end{aligned}$$

Check the solution in the original equation.

$$\begin{aligned} 2x-3&=-5y \\ 2\left(\frac{11}{19}\right)-3&\stackrel{?}{=}-5\left(\frac{7}{19}\right) \\ \frac{22}{19}-\frac{57}{19}&\stackrel{?}{=}-\frac{35}{19} \\ -\frac{35}{19}&=-\frac{35}{19}\ \checkmark \end{aligned}$$

$$\begin{aligned} -2y&=-3x+1 \\ -2\left(\frac{7}{19}\right)&\stackrel{?}{=}-3\left(\frac{11}{19}\right)+1 \\ -\frac{14}{19}&\stackrel{?}{=}-\frac{33}{19}+\frac{19}{19} \\ -\frac{14}{19}&=-\frac{14}{19}\ \checkmark \end{aligned}$$

These are both true statements. The solution is $x=\frac{11}{19}$, $y=\frac{7}{19}$; this is also written as the point $\left(\frac{11}{19}, \frac{7}{19}\right)$.

If the elimination method produces a sentence that is always true, then the system is dependent, and values that satisfy either original equation are solutions. If the elimination method produces a sentence that is always false, then the system is inconsistent, and there is no solution.

Solving Systems of Equations Using Matrices

A **matrix** (plural, **matrices**) is a rectangular array of numbers or variables. A matrix can be used to represent a system of equations in standard form by writing only the coefficients of the variables and the constants from the equations.

Example 1: Represent this system as a matrix.

$$\begin{cases} 5x - 2y = 13 \\ 2x + y = 7 \end{cases}$$

$$\left.\begin{matrix} 5x - 2y = 13 \\ 2x + y = 7 \end{matrix}\right\} \rightarrow \begin{bmatrix} 5 & -2 & \vdots & 13 \\ 2 & 1 & \vdots & 7 \end{bmatrix}$$

In the preceding matrix, the dashed line separates the coefficients of the variables from the constants in each equation.

Through the use of row multiplication and row additions, the goal is to transform the preceding matrix into the following form.

$$\begin{bmatrix} a & b & \vdots & c \\ 0 & d & \vdots & e \end{bmatrix}$$

The matrix method is the same as the elimination method but more organized.

Example 2: Solve this system by using matrices.

$$\begin{cases} 5x - 2y = 13 \\ 2x + y = 7 \end{cases}$$

$$\left.\begin{matrix} 5x - 2y = 13 \\ 2x + y = 7 \end{matrix}\right\} \rightarrow \begin{bmatrix} 5 & -2 & \vdots & 13 \\ 2 & 1 & \vdots & 7 \end{bmatrix} \begin{matrix} \text{multiply (2)} \\ \text{multiply (–5)} \end{matrix}$$

Multiply row 1 by 2 and multiply row 2 by –5, then add the columns:

$$\left[\begin{array}{rr:r} 10 & -4 & 26 \\ -10 & -5 & -35 \\ \hline 0 & -9 & -9 \end{array}\right]$$

$$\begin{array}{r} \text{Rewrite row 1} \rightarrow \\ \text{Replace that sum in the original row 2} \rightarrow \end{array} \left[\begin{array}{rr:r} 5 & -2 & 13 \\ 0 & -9 & -9 \end{array}\right]$$

This matrix now represents the system

$$\begin{aligned} 5x - 2y &= 13 \\ -9y &= -9 \end{aligned}$$

Therefore, $y = 1$.

Now, substitute 1 for y in the other equation and solve for x.

$$\begin{aligned} 5x - 2y &= 13 \\ 5x - 2(1) &= 13 \\ 5x &= 15 \\ x &= 3 \end{aligned}$$

Check the solution.

$$\begin{aligned} 5x - 2y &= 13 \\ 5(3) - 2(1) &\stackrel{?}{=} 13 \\ 15 - 2 &\stackrel{?}{=} 13 \\ 13 &= 13 \checkmark \end{aligned} \qquad \begin{aligned} 2x + y &= 7 \\ 2(3) + 1 &\stackrel{?}{=} 7 \\ 6 + 1 &\stackrel{?}{=} 7 \\ 7 &= 7 \checkmark \end{aligned}$$

The solution is $x = 3$, $y = 1$; this is also written as the point (3, 1).

Matrices are an alternative method of solving linear systems of equations. Matrices are a more time-consuming method than either the elimination or substitution method. They only become a time-saving method when solving multiple equations in multiple variables.

Solving Systems of Equations Using Determinants (Cramer's Rule)

A square array of numbers or variables enclosed between vertical lines is called a **determinant.** A determinant can characterize a matrix, but it is not one itself. A determinant has a numerical value, whereas a matrix does not. The following determinant has two rows and two columns.

$$\begin{array}{rc} \text{row } 1 \rightarrow & \left| \begin{array}{cc} a & c \\ b & d \end{array} \right| \\ & \begin{array}{cc} \uparrow & \uparrow \\ \text{column} & \text{column} \\ 1 & 2 \end{array} \end{array}$$

(row 2 → points to the second row)

The value of this determinant is found by finding the difference between the diagonally down product and the diagonally up product:

$$\begin{vmatrix} a & c \\ b & d \end{vmatrix} = ad - bc$$

Example 1: Evaluate the following determinant.

$$\begin{vmatrix} 3 & -11 \\ 7 & 2 \end{vmatrix}$$

$$\begin{vmatrix} 3 & -11 \\ 7 & 2 \end{vmatrix} = (3)(2) - (7)(-11)$$
$$= 6 - (-77)$$
$$= 83$$

Finding solutions of systems by using determinants is referred to as **Cramer's Rule,** named after the mathematician who devised this method. Cramer's Rule could hardly be considered a "shortcut," but it is a valid way to solve systems of equations by using determinants.

Here are the steps for solving a two-variable system using Cramer's Rule:

1. Create the determinant D by using the x-coefficients and the y-coefficients from the system. The original equations need to be written in standard form.
2. Find the value of that determinant and call it D.
3. Create the determinant D_x. The x-numerator determinant is formed by taking the constant terms from the system and placing them in the x-coefficient positions and retaining the y-coefficients.
4. Find the value of that determinant.
5. Create the determinant D_y. The y-numerator determinant is formed by taking the constant terms from the system and placing them in the y-coefficient positions and retaining the x-coefficients.
6. Find the value of that determinant.

The solution to the system is the ordered pair (x, y), where $x=\frac{D_x}{D}$ and $y=\frac{D_y}{D}$.

Example 2: Solve the following system by using determinants.

$$\begin{cases} 4x-3y=-14 \\ 3x-5y=-5 \end{cases}$$

To solve this system, the three determinants are created. One is called the denominator determinant, labeled D; another is the x-numerator determinant, labeled D_x; and the third is the y-numerator determinant, labeled D_y.

$$\begin{aligned} D &= \begin{vmatrix} 4 & -3 \\ 3 & -5 \end{vmatrix} \\ &= (4)(-5)-(3)(-3) \\ &= -20-(-9) \\ &= -20+9 \\ &= -11 \end{aligned}$$

$$\begin{aligned} D_x &= \begin{vmatrix} -14 & -3 \\ -5 & -5 \end{vmatrix} \\ &= (-14)(-5) - (-5)(-3) \\ &= 70 - 15 \\ &= 55 \end{aligned}$$

$$\begin{aligned} D_y &= \begin{vmatrix} 4 & -14 \\ 3 & -5 \end{vmatrix} \\ &= (4)(-5) - (3)(-14) \\ &= -20 - (-42) \\ &= -20 + 42 \\ &= 22 \end{aligned}$$

The answers for x and y are as follows:

$$x = \frac{D_x}{D} = \frac{55}{-11} = -5 \qquad y = \frac{D_y}{D} = \frac{22}{-11} = -2$$

The solution is $x = -5, y = -2$; this is also written as the point $(-5, -2)$.

Example 3: Use Cramer's Rule to solve this system.

$$\begin{cases} 4x + 6y = 3 \\ 8x - 3y = 1 \end{cases}$$

$$D = \begin{vmatrix} 4 & 6 \\ 8 & -3 \end{vmatrix} = -12 - 48 = -60$$

$$D_x = \begin{vmatrix} 3 & 6 \\ 1 & -3 \end{vmatrix} = -9 - 6 = -15$$

$$D_y = \begin{vmatrix} 4 & 3 \\ 8 & 1 \end{vmatrix} = 4 - 24 = -20$$

$$x = \frac{D_x}{D} = \frac{-15}{-60} = \frac{1}{4}, \; y = \frac{D_y}{D} = \frac{-20}{-60} = \frac{1}{3}$$

The solution is $x=\frac{1}{4}$, $y=\frac{1}{3}$. Also written $\left(\frac{1}{4}, \frac{1}{3}\right)$.

Solving Word Problems That Translate into Linear Systems

Many word problems can be translated into linear systems. The solution to the system can be found in the various ways previously studied. Interpreting the solution to the system as an answer to the word problem becomes the final step.

Example 1: Joy wants to invest a total of \$5,000 in two different annuities. One annuity pays 5% annually and the other pays 4% annually. How much should Joy invest in each annuity for her annual annuity income to be \$235?

Use x to represent the money invested at 5% and y to represent the money invested at 4%. Their total sum is \$5,000; 5% of the first and 4% of the second sums to 235. The system becomes

$$\begin{cases} x+y=5{,}000 \\ 0.05x+0.04y=235 \end{cases}$$

$$\begin{array}{ll} x+y=5{,}000 & \\ \underline{0.05x+0.04y=235} & \text{Multiply by 100} \end{array}$$

$$\begin{array}{ll} x+y=\ 5{,}000 & \text{Multiply by} -5 \\ \underline{5x+4y=23{,}500} & \end{array}$$

$$\begin{array}{r} -5x-5y=-25{,}000 \\ \underline{5x+4y=\ 23{,}500} \\ -y=\ -1{,}500 \end{array}$$

$$y=\$1{,}500 \text{ invested at } 4\%$$

Substitute $y = \$1{,}500$ into the first equation.

$$x+y=5{,}000$$
$$x+1{,}500=5{,}000$$
$$x=5{,}000-1{,}500=\$3{,}500 \text{ invested at } 5\%$$

Joy should invest \$1,500 at 4% and \$3,500 at 5%.

Example 2: Ed has 50 coins in his hands, all quarters and dimes. The total value of his coins is \$9.80. How many of each kind of coin is Ed holding?

Using x to represent the number of dimes and y to represent the number of quarters, the system becomes

$$\begin{cases} x+y=50 \\ 0.10x+0.25y=9.80 \end{cases}$$

$$\begin{aligned} x+y&=50 \\ 0.10x+0.25y&=9.80 \quad \text{Multiply by 100} \end{aligned}$$

$$\begin{aligned} x+y&=50 \quad \text{Multiply by} -10 \\ 10x+25y&=980 \end{aligned}$$

$$\begin{aligned} -10x-10y&=-500 \\ 10x+25y&=\ \ 980 \\ 15y&=\ \ 480 \\ y&=32 \text{ quarters} \end{aligned}$$

Substitute $y = 32$ into the first equation.

$$\begin{aligned} x+y&=50 \\ x+32&=50 \\ x&=50-32 \\ x&=18 \text{ dimes} \end{aligned}$$

Ed has 32 quarters and 18 dimes.

Example 3: A chemist has an alloy of copper and silver that is 25% silver and another alloy of copper and silver that is 37.5% silver. How much of each alloy should he use to produce 200 pounds of an alloy that is 30% silver?

Using x to represent the amount of the 25% alloy and y to represent the amount of the 37.5% alloy, the system becomes

$$\begin{cases} x+y=200 \\ 0.25x+0.375y=0.30(200) \end{cases}$$

$$\begin{aligned} x+y&=200 \\ 0.25x+0.375y&=0.30(200) \quad \text{Multiply by 1,000} \end{aligned}$$

$$\begin{aligned} x+y&=200 \quad \text{Multiply by} -250 \\ 250x+375y&=60{,}000 \end{aligned}$$

$$\begin{aligned} -250x-250y&=-50{,}000 \\ 250x+375y&=\ \ 60{,}000 \\ 125y&=\ \ 10{,}000 \\ y&=80 \text{ pounds of 37.5\% alloy} \end{aligned}$$

Substitute $y = 80$ into the first equation.

$$\begin{aligned} x+y&=200 \\ x+80&=200 \\ x&=200-80=120 \text{ pounds of 25\% alloy} \end{aligned}$$

The chemist should use 80 pounds of the 37.5% alloy and120 pounds of the 25% alloy.

Example 4: An airplane can fly 3,000 miles into a headwind in 6 hours. With the same wind blowing, that plane can fly the same distance in the opposite direction in an hour less time. Find the rate of speed of the plane and the rate of speed of the wind.

Use x to represent the speed of the airplane and y to represent the speed of the wind. We can add or subtract the values to find the speed of the plane going with or against the wind. The system becomes

$$\begin{cases} 6(x-y)=3{,}000 \\ 5(x+y)=3{,}000 \end{cases}$$

$$\begin{array}{ll} 6(x-y)=3{,}000 & \text{Divide by 6} \\ \underline{5(x+y)=3{,}000} & \text{Divide by 5} \end{array}$$

$$\begin{aligned} x-y&=500 \\ \underline{x+y}&\underline{=600} \\ 2x\quad &=1{,}100 \\ x&=550 \text{ mph speed of the airplane} \end{aligned}$$

Substitute $x = 550$ into the first equation.

$$\begin{aligned} 6(x-y)&=3{,}000 \\ 6(550-y)&=3{,}000 \\ 3{,}300-6y&=3{,}000 \\ -6y&=-300 \\ y&=50 \text{ mph speed of the wind} \end{aligned}$$

The speed of the airplane is 550 mph, and the speed of the wind is 50 mph.

Example 5: The sum of the measures of the five angles in a pentagon is 540°. In pentagon *FGHIJ*, angles *F*, *G*, and *H* are congruent, and angles *I* and *J* are congruent. The measure of angle *G* is 30° less than the measure of angle *J*. Find the measure of each angle.

Using x to represent the measure of each angle *F*, *G*, and *H*, and using y to represent the measure of each angle *I* and *J*, the system becomes

$$\begin{cases} 3x+2y=540 \\ \quad\quad x=y-30 \end{cases}$$

Substitute $y - 30$ for x in first equation.

$$\begin{aligned} 3x+2y&=540 \\ 3(y-30)+2y&=540 \\ 3y-90+2y&=540 \\ 5y&=630 \\ y&=126^\circ, \text{ the measure of each angle } I \text{ and } J \end{aligned}$$

Substitute $y = 126$ into second equation.

$$\begin{aligned} x&=y-30 \\ x&=126-30 \\ x&=96^\circ, \text{ the measure of each angle } F,\ G, \text{ and } H \end{aligned}$$

Therefore, angles I and J each measure 126° and angles F, G, and H each measure 96°.

Solving Systems of Linear Inequalities by Graphing

To graph the solution to a system of inequalities, follow this procedure:

1. Graph each sentence on the same set of axes.
2. Determine where the shading of the sentences overlaps. The overlapping region is the solution to the system of inequalities.

Example 1: Graph the solution to this system of inequalities.

$$\begin{cases} x+2y \le 12 \\ 3x-y \ge 6 \end{cases}$$

The solution to this system is the region with both shadings, as shown in the following figure.

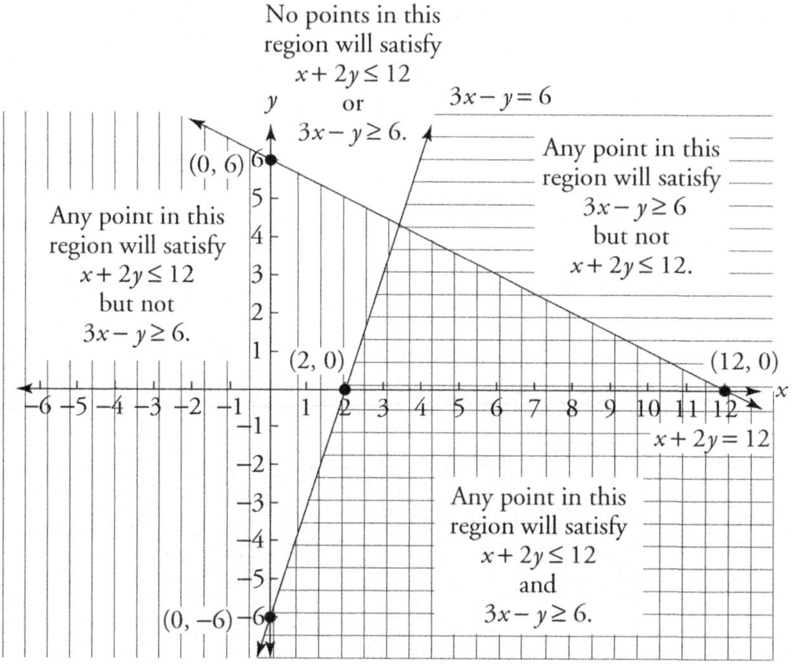

Here is the final graph of the solution to this system of inequalities.

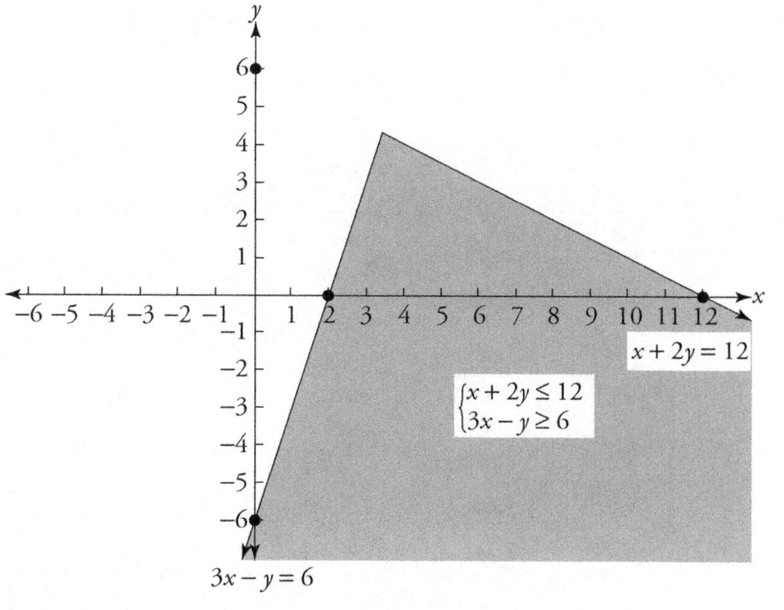

Example 2: Graph the solution to this system of inequalities.

$$\begin{cases} y > x + 2 \\ y < x - 1 \end{cases}$$

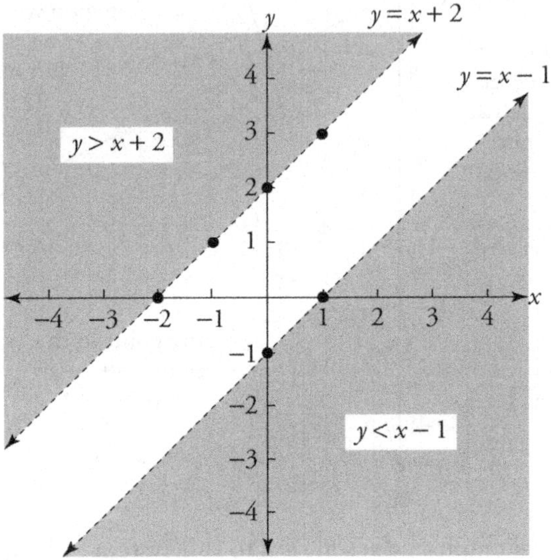

Because no overlapping regions exist, this system has no solutions.

Example 3: Solve this system by graphing.

$$\begin{cases} x \geq 2 \\ y \leq 4 \\ x - y < 6 \end{cases}$$

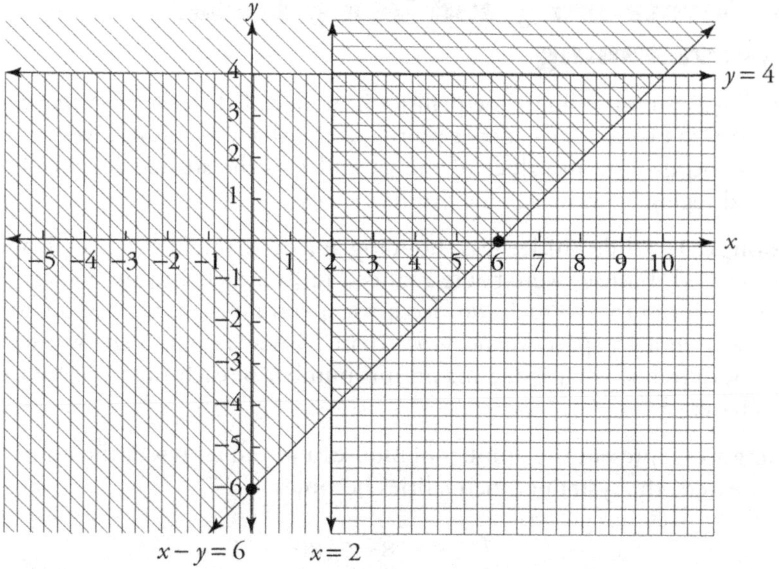

Here is the final graph of the solution to this system of inequalities.

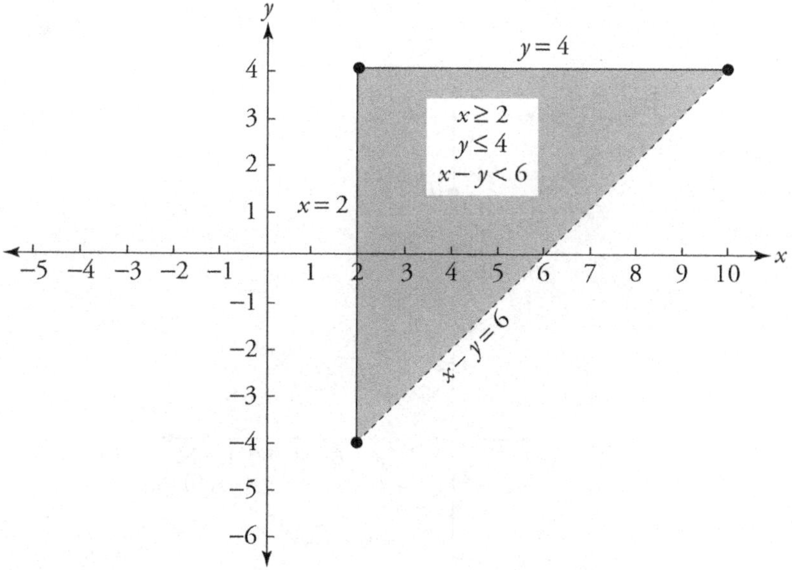

Understanding and Using Linear Programming

Solving a system of linear inequalities is often called *linear programming*. Finding the vertices, or corner points of the shaded region, gives maximum and minimum values of the variables in the shaded region. The shaded region is known as the *feasibility region.*

Example 1: While Shirlee was shopping at a department store, a sale was announced. Blouses were being reduced to \$12 each and sweaters were being reduced to \$18 each. Shirlee must spend no more than \$180, and has only enough space for no more than 13 new items. If she wants to buy both items, what is the maximum number of each item Shirlee can purchase?

Using x to represent the number of blouses and y to represent the number of sweaters, the system of inequalities becomes

$$\begin{cases} 12x+18y \le 180 \\ x+y \le 13 \end{cases}$$

Graph both inequalities. The x and y values are restricted to Quadrant I since both the blouses and the sweaters represent positive values.

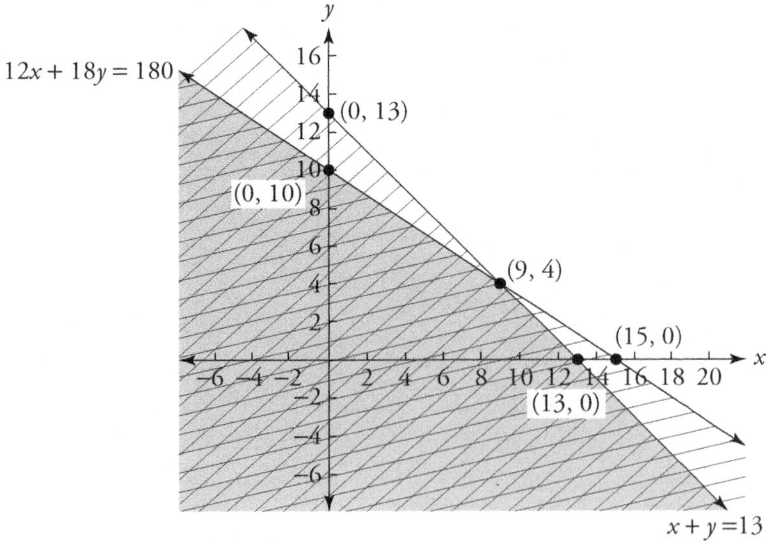

The corner points of the shaded feasibility region in Quadrant I are (0, 10), (13, 0), and (9, 4). At these points, the most (or least) number of items may have been purchased.

The point (9, 4) represents the intersection of the two lines.

Solve for x and y.

$$\begin{aligned} 12x + 18y &= 180 \\ x + \quad y &= \ 13 \quad \text{Multiply by } -12 \end{aligned}$$

$$\begin{aligned} 12x + 18y &= \ \ 180 \\ -12x - 12y &= -156 \\ \hline 6y &= \ \ 24 \\ y &= 4 \end{aligned}$$

Substitute $y = 4$ into the second equation.

$$\begin{aligned} x + y &= 13 \\ x + 4 &= 13 \\ x &= 13 - 4 \\ x &= 9 \end{aligned}$$

The point (9, 4) is one of the three corner points.

Therefore, the maximum number of each item that Shirlee can purchase is 4 sweaters and 9 blouses.

Chapter Check-Out

Questions

1. Solve by substitution: $\begin{cases} 5x+3y=30 \\ 6y-4x=18 \end{cases}$.

2. Solve by elimination: $\begin{cases} 4x+7y=56 \\ 9x-14y=7 \end{cases}$.

3. Evaluate the determinant: $\begin{vmatrix} 4 & -8 \\ 6 & 7 \end{vmatrix}$.

4. Use Cramer's Rule to solve this system: $\begin{cases} 2x+3y=39 \\ 8x-5y=3 \end{cases}$.

5. Describe what it means to solve a system of linear equations by graphing.

6. Explain the difference between a matrix and a determinant.

7. Spencer likes Belgian waffles and cranberry juice for breakfast. He needs to eat less than or equal to 500 calories and no more than 425 mg of sodium. The Belgian waffles have 90 calories each and 100 mg of sodium each. The cranberry juice has 140 calories and 25 mg of sodium per small-glass serving. If Spencer wants to maximize his calorie intake and milligrams of sodium in this meal, find the maximum number of Belgian waffles and small-glass servings of cranberry juice that Spencer can have in this breakfast. ***Hint:*** Write an inequality for the calories and an inequality for the sodium. The graphs are limited to Quadrant I. Use x to represent the number of Belgian waffles. Use y to represent the small-glass servings of cranberry juice. Show your work.

Answers

1. $x = 3, y = 5$
2. $x = 7, y = 4$
3. 76
4. $x = 6, y = 9$
5. The solution is represented by the point (or points)—if it exists—where the two lines intersect.
6. A matrix has no numerical value; a determinant has a numerical value. A matrix is formed using coefficients and/or the constant terms from both equations. A determinant is formed using coefficients or constant terms.
7. The system becomes $\begin{cases} 90x + 140y \le 500 \\ 100x + 25y \le 425 \end{cases}$.

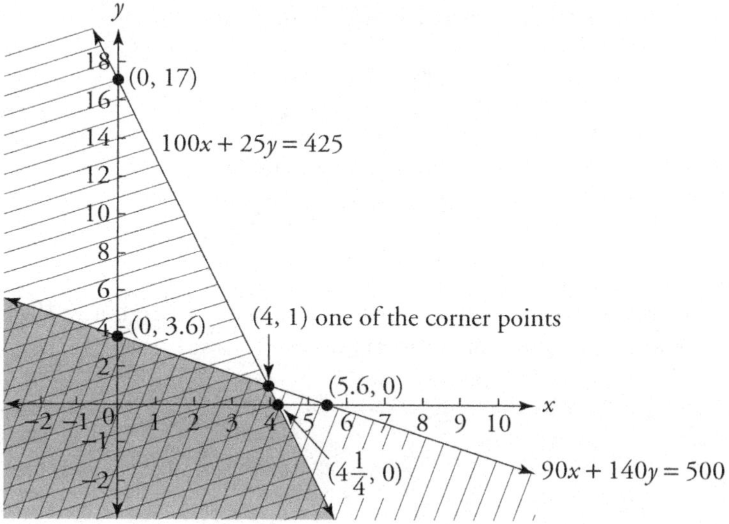

Solve for x and y.

$$\begin{array}{rl} 90x+140y=500 & \text{Divide by 10} \\ 100x+\ \ 25y=425 & \text{Divide by 25} \\ \hline \end{array}$$

$$\begin{array}{rl} 9x+14y=50 & \\ 4x+\quad y=17 & \text{Multiply by } -14 \\ \hline \end{array}$$

$$\begin{array}{r} 9x+14y=\quad 50 \\ -56x-14y=-238 \\ \hline -47x\qquad\quad =-188 \\ x=4 \end{array}$$

Substitute $x = 4$ into the first equation.

$$\begin{aligned} 90x+140y&=500 \\ 90(4)+140y&=500 \\ 360+140y&=500 \\ 140y&=140 \\ y&=1 \end{aligned}$$

The corner point is (4, 1). Spencer can have 4 Belgian waffles and 1 small-glass serving of cranberry juice.

Chapter 4

LINEAR SYSTEMS IN THREE VARIABLES

Chapter Check-In

- ❑ Solving three-variable systems of equations by elimination
- ❑ Solving three-variable systems of equations using matrices
- ❑ Solving three-variable systems of equations using determinants and Cramer's Rule
- ❑ Translating word problems into three-variable systems

Common Core Standard: Seeing the Structure in Expressions

Solving equations and inequalities in three variables using elimination, matrices, and determinants. Reasoning with equations and inequalities (A.REI).

Systems of equations with three variables are only slightly more complicated to solve than those with two variables (covered in Chapter 3). Three-variable systems may contain three different variables and three different equations. The solution is referred to as an *ordered triple*. It is written in the form of (x, y, z), where the values of the variables are written in alphabetical order of the variables. The standard form of a three-variable equation is $ax + by + cz = d$, where the coefficients and the constant terms are real numbers. The two most straightforward methods of solving these types of equations are by elimination and by using 3×3 matrices. Both are covered extensively in this chapter. This chapter also further explores Cramer's Rule—the application of using determinants to solve systems.

Solving Three-Variable Systems of Equations by Elimination

To use elimination to solve a system of three equations with three variables, follow this procedure:

1. Write all the equations in standard form cleared of decimals or fractions.
2. Choose a variable to eliminate, then choose any two of the three equations and eliminate the chosen variable.
3. Select a different set of two equations and eliminate the same variable as in Step 2.
4. Solve the two equations from steps 2 and 3 for the two variables they contain.
5. Substitute the answers from Step 4 into any equation involving the remaining variable.
6. Check the solution with all three original equations.

Example 1: Solve this system of equations using elimination.

$$\begin{cases} 4x-2y+3z=1 & (1) \\ x+3y-4z=-7 & (2) \\ 3x+y+2z=5 & (3) \end{cases}$$

All the equations are already in the required form.

Choose a variable to eliminate, say *x*, and select two equations with which to eliminate it, say equations (1) and (2).

$$\begin{array}{lrcr} & 4x-2y+3z=1 & \rightarrow & 4x-2y+3z=1 \\ \text{Multiply by } -4 & x+3y-4z=-7 & \rightarrow & \underline{-4x-12y+16z=28} \\ & & & -14y+19z=29 \quad (4) \end{array}$$

Select a different set of two equations, say equations (2) and (3), and eliminate the same variable.

$$\begin{array}{lrcr} \text{Multiply by } -3 & x+3y-4z=-7 & \rightarrow & -3x-9y+12z=21 \\ & 3x+y+2z=5 & \rightarrow & \underline{3x+y+2z=5} \\ & & & -8y+14z=26 \quad (5) \end{array}$$

Solve the system created by equations (4) and (5).

$$\begin{array}{lrcr} \text{Multiply by } -8 & -14y+19z=29 \rightarrow & 112y-152z= & -232 \\ \text{Multiply by } 14 & -8y+14z=26 \rightarrow & -112y+196z= & 364 \\ \hline & & 44z= & 132 \\ & & z= & 3 \end{array}$$

Now, substitute $z = 3$ into equation (4) to find y.

$$\begin{aligned} -14y+19z&=29 \\ -14y+19(3)&=29 \\ -14y+57&=29 \\ -14y&=-28 \\ y&=2 \end{aligned}$$

Use the answers from Step 4 and substitute into any equation involving the remaining variable.

Using equation (2),

$$\begin{aligned} x+3y-4z&=-7 \\ x+3(2)-4(3)&=-7 \\ x+6-12&=-7 \\ x&=-1 \end{aligned}$$

Check the solution in all three original equations.

$$\begin{aligned} 4x-2y+3z&=1 \\ 4(-1)-2(2)+3(3)&\stackrel{?}{=}1 \\ -4-4+9&\stackrel{?}{=}1 \\ 1&=1 \checkmark \end{aligned}$$

$$\begin{aligned} x+3y-4z&=-7 \\ -1+3(2)-4(3)&\stackrel{?}{=}-7 \\ -1+6-12&\stackrel{?}{=}-7 \\ -7&=-7 \checkmark \end{aligned}$$

$$
\begin{aligned}
3x+y+2z&=5\\
3(-1)+2+2(3)&\stackrel{?}{=}5\\
-3+2+6&\stackrel{?}{=}5\\
5&=5\ \checkmark
\end{aligned}
$$

The solution is $x = -1, y = 2, z = 3$; this is also written $(-1, 2, 3)$.

Example 2: Solve this system of equations using the elimination method.

$$
\begin{cases}
x=3z-5 & (1)\\
2x+2z=y+16 & (2)\\
7x-5z=3y+19 & (3)
\end{cases}
$$

Write all equations in standard form.

$$
\begin{cases}
x-3z=-5 & (1)\\
2x-y+2z=16 & (2)\\
7x-3y-5z=19 & (3)
\end{cases}
$$

Notice that equation (1) already has the y eliminated. Therefore, use equations (2) and (3) to eliminate y. Then use this result, together with equation (1), to solve for x and z. Use these results and substitute into either equation (2) or (3) to find y.

$$
\begin{array}{lrcr}
\text{Multiply by} -3 & 2x-\ \ y+2z=16 & \rightarrow & -6x+3y-6z=-48\\
 & 7x-3y-5z=19 & \rightarrow & 7x-3y-5z=\ \ 19\\
\hline
 & & & x\qquad -11z=-29
\end{array}
$$

$$
\begin{array}{lrcr}
\text{Multiply by} -1 & x-\ \ 3z=\ -5 & \rightarrow & -x+\ 3z=\quad 5\\
 & x-11z=-29 & \rightarrow & x-11z=-29\\
\hline
 & & & -8z=-24\\
 & & & z=3
\end{array}
$$

Substitute $z = 3$ into equation (1).

$$\begin{aligned} x-3z&=-5 \\ x-3(3)&=-5 \\ x-9&=-5 \\ x&=4 \end{aligned}$$

Substitute $x = 4$ and $z = 3$ into equation (2).

$$\begin{aligned} 2x-y+2z&=16 \\ 2(4)-y+2(3)&=16 \\ 8-y+6&=16 \\ -y&=2 \\ y&=-2 \end{aligned}$$

Use the original equations to check the solution. The solution is $x = 4$, $y = -2$, $z = 3$; this is also written as $(4, -2, 3)$.

Solving Three-Variable Systems of Equations Using Matrices

As explained in Chapter 3, solving a system of equations by using matrices is merely an organized manner of using the elimination method.

Example 1: Solve the following system of equations using matrices.

$$\begin{cases} 4x+9y=8 \\ 8x+6z=-1 \\ 6y+6z=-1 \end{cases}$$

Put the equations in matrix form.

$$\left.\begin{aligned} 4x+9y&=8 \\ 8x+6z&=-1 \\ 6y+6z&=-1 \end{aligned}\right\} \rightarrow \begin{bmatrix} 4 & 9 & 0 & : & 8 \\ 8 & 0 & 6 & : & -1 \\ 0 & 6 & 6 & : & -1 \end{bmatrix} \begin{matrix} (1) \\ (2) \\ (3) \end{matrix}$$

Eliminate the *x*-coefficient below row 1.

$$\begin{array}{r} \text{Retain row (1)} \\ \text{Replace row (2) with row (3)} \\ \text{Add } -2 \text{ times row (1) plus 1 times row (2)} \end{array} \left[\begin{array}{ccc:c} 4 & 9 & 0 & 8 \\ 0 & 6 & 6 & -1 \\ 0 & -18 & 6 & -17 \end{array}\right] \begin{array}{c} (4) \\ (5) \\ (6) \end{array}$$

Eliminate the *y*-coefficient below row 5.

$$\begin{array}{r} \text{Retain row (4)} \\ \text{Retain row (5)} \\ \text{Add 3 times row (5) plus 1 times row (6)} \end{array} \left[\begin{array}{ccc:c} 4 & 9 & 0 & 8 \\ 0 & 6 & 6 & -1 \\ 0 & 0 & 24 & -20 \end{array}\right] \begin{array}{c} (7) \\ (8) \\ (9) \end{array}$$

Reinserting the variables, the system is now:

$$\begin{aligned} 4x+9y&=8 && (7) \\ 6y+6z&=-1 && (8) \\ 24z&=-20 && (9) \end{aligned}$$

Equation (9) can be solved for *z*.

$$\begin{aligned} 24z&=-20 \\ z&=-\frac{20}{24} \\ z&=-\frac{5}{6} \end{aligned}$$

Substitute $z=-\frac{5}{6}$ into equation (8) and solve for y.

$$\begin{aligned}6y+6z&=-1\\6y+6\left(-\frac{5}{6}\right)&=-1\\6y-5&=-1\\6y&=4\\y&=\frac{4}{6}\\y&=\frac{2}{3}\end{aligned}$$

Substitute $y=\frac{2}{3}$ into equation (7) and solve for x.

$$\begin{aligned}4x+9y&=8\\4x+9\left(\frac{2}{3}\right)&=8\\4x+6&=8\\4x&=2\\x&=\frac{2}{4}\\x&=\frac{1}{2}\end{aligned}$$

The solution is $x=\frac{1}{2}$, $y=\frac{2}{3}$, $z=-\frac{5}{6}$; this is also written as $\left(\frac{1}{2}, \frac{2}{3}, -\frac{5}{6}\right)$.

Example 2: Solve this system of equations by using matrices.

$$\begin{cases}2x+y-3z=-4\\4x-2y+z=9\\3x+5y-2z=5\end{cases}$$

The goal is to arrive at a matrix of the following form.

$$\begin{bmatrix} a & b & c & : & d \\ 0 & e & f & : & g \\ 0 & 0 & h & : & i \end{bmatrix}$$

To do this, you use row multiplications, row additions, or interchanging rows, as shown in the following.

Put the equation in matrix form.

$$\begin{bmatrix} 2 & 1 & -3 & : & -4 \\ 4 & -2 & 1 & : & 9 \\ 3 & 5 & -2 & : & 5 \end{bmatrix} -2R_1 + R_2 \rightarrow R_2 \begin{bmatrix} 2 & 1 & -3 & : & -4 \\ 0 & -4 & 7 & : & 17 \\ 3 & 5 & -2 & : & 5 \end{bmatrix} -3R_1 + 2R_3 \rightarrow R_3$$

$$\begin{bmatrix} 2 & 1 & -3 & : & -4 \\ 0 & -4 & 7 & : & 17 \\ 0 & 7 & 5 & : & 22 \end{bmatrix} 7R_2 + 4R_3 \rightarrow R_3 \begin{bmatrix} 2 & 1 & -3 & : & -4 \\ 0 & -4 & 7 & : & 17 \\ 0 & 0 & 69 & : & 207 \end{bmatrix}$$

From row 3

$$69z = 207$$
$$z = 3$$

Substitute $z = 3$ into row 2.

$$-4y + 7z = 17$$
$$-4y + 7(3) = 17$$
$$-4y + 21 = 17$$
$$-4y = -4$$
$$y = 1$$

Substitute $z = 3$ and $y = 1$ into row 1.

$$
\begin{aligned}
2x + y - 3z &= -4\\
2x + 1 - 3(3) &= -4\\
2x + 1 - 9 &= -4\\
2x - 8 &= -4\\
2x &= 4\\
x &= 2
\end{aligned}
$$

The solution is (2, 1, 3).

Solving Three-Variable Systems Using Determinants (Cramer's Rule)

Evaluating determinants and using determinants to solve three-variable systems is a process known as Cramer's Rule. Evaluating determinants was explained in Chapter 3. The determinant of a 2×2 matrix is defined as follows:

$$\begin{vmatrix} a & c \\ b & d \end{vmatrix} = ad - bc$$

The determinant of a 3×3 matrix can be defined as shown in the following.

minor determinants

$$\begin{vmatrix} a_1 & b_1 & c_1 \\ a_2 & b_2 & c_2 \\ a_3 & b_3 & c_3 \end{vmatrix} = a_1\begin{vmatrix} b_2 & c_2 \\ b_3 & c_3 \end{vmatrix} - a_2\begin{vmatrix} b_1 & c_1 \\ b_3 & c_3 \end{vmatrix} + a_3\begin{vmatrix} b_1 & c_1 \\ b_2 & c_2 \end{vmatrix}$$

subtract add

$$= a_1(b_2c_3 - b_3c_2) - a_2(b_1c_3 - b_3c_1) + a_3(b_1c_2 - b_2c_1)$$

This process is known as *expanding by minors.* Each minor determinant is obtained by crossing out the first column and one row.

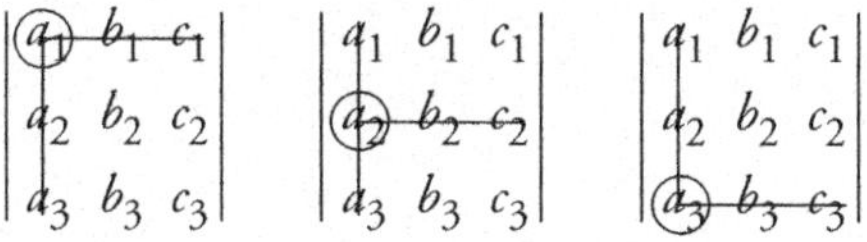

Example 1: Evaluate the following determinant.

$$\begin{vmatrix} -2 & 4 & 1 \\ -3 & 6 & -2 \\ 4 & 0 & 5 \end{vmatrix}$$

First, find the minor determinants.

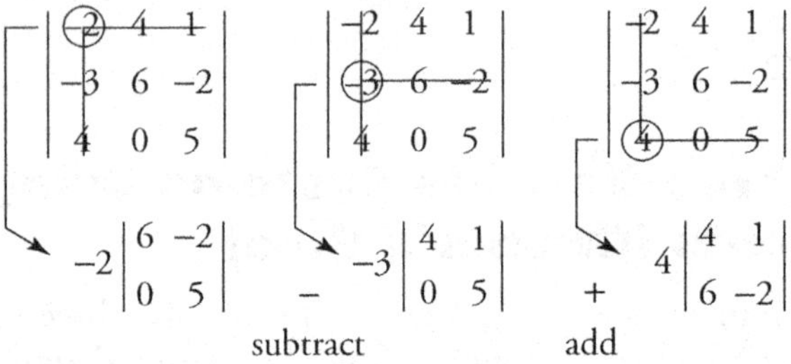

$$-2(30-0) \quad - \quad -3(20-0) \quad + \quad 4(-8-6)$$

$$-60 \quad - \quad -60 \quad + \quad -56$$

$$-60 \quad +60 \quad -56 \quad = -56$$

The solution is

$$\begin{vmatrix} -2 & 4 & 1 \\ -3 & 6 & -2 \\ 4 & 0 & 5 \end{vmatrix} = -56$$

To use determinants to solve a system of three equations with three variables—say x, y, and z—four determinants must be formed using the following steps. This procedure is known as Cramer's Rule.

1. Write all equations in standard form.
2. Create the denominator determinant, D, by using the coefficients of x, y, and z from the original equations and evaluate it.

3. Create the x-numerator determinant, D_x, the y-numerator determinant, D_y, and the z-numerator determinant, D_z, by replacing the respective x, y, and z coefficients with the constants from the original equations in standard form and evaluate each determinant.

The answers for x, y, and z are as follows:

$$x=\frac{D_x}{D},\quad y=\frac{D_y}{D},\quad z=\frac{D_z}{D}$$

Example 2: Solve this system of equations using Cramer's Rule.

$$\begin{cases}3x+2y-z=2\\2x-y-3z=13\\x+3y-2z=1\end{cases}$$

Find the minor determinants.

x-coefficients ↓ *y*-coefficients ↓ *z*-coefficients ↓

$$D=\begin{vmatrix}3 & 2 & -1\\2 & -1 & -3\\1 & 3 & -2\end{vmatrix}=3\begin{vmatrix}-1 & -3\\3 & -2\end{vmatrix}-2\begin{vmatrix}2 & -1\\3 & -2\end{vmatrix}+1\begin{vmatrix}2 & -1\\-1 & -3\end{vmatrix}$$

$$\begin{aligned}&=3[2-(-9)]-2[-4-(-3)]+1(-6-1)\\&=3(11)-2(-1)+1(-7)\\&=33+2-7=28\end{aligned}$$

Use the constants to replace the *x*-coefficients.

constants replacing the *x*-coefficients ↓

$$D_x = \begin{vmatrix} 2 & 2 & -1 \\ 13 & -1 & -3 \\ 1 & 3 & -2 \end{vmatrix} = 2\begin{vmatrix} -1 & -3 \\ 3 & -2 \end{vmatrix} - 13\begin{vmatrix} 2 & -1 \\ 3 & -2 \end{vmatrix} + 1\begin{vmatrix} 2 & -1 \\ -1 & -3 \end{vmatrix}$$

$$= 2[2-(-9)] - 13[-4-(-3)] + 1(-6-1)$$

$$= 2(11) - 13(-1) + 1(-7)$$

$$= 22 + 13 - 7 = 28$$

Use the constants to replace the *y*-coefficients.

constants replacing the *y*-coefficients ↓

$$D_y = \begin{vmatrix} 3 & 2 & -1 \\ 2 & 13 & -3 \\ 1 & 1 & -2 \end{vmatrix} = 3\begin{vmatrix} 13 & -3 \\ 1 & -2 \end{vmatrix} - 2\begin{vmatrix} 2 & -1 \\ 1 & -2 \end{vmatrix} + 1\begin{vmatrix} 2 & -1 \\ 13 & -3 \end{vmatrix}$$

$$= 3[-26-(-3)] - 2[-4-(-1)] + 1[-6-(-13)]$$

$$= 3(-23) - 2(-3) + 1(7)$$

$$= -69 + 6 + 7 = -56$$

Use the constants to replace the *z*-coefficients.

constants replacing the *z*-coefficients

$$D_z = \begin{vmatrix} 3 & 2 & 2 \\ 2 & -1 & 13 \\ 1 & 3 & 1 \end{vmatrix} = 3\begin{vmatrix} -1 & 13 \\ 3 & 1 \end{vmatrix} - 2\begin{vmatrix} 2 & 2 \\ 3 & 1 \end{vmatrix} + 1\begin{vmatrix} 2 & 2 \\ -1 & 13 \end{vmatrix}$$

$$= 3(-1-39) - 2(2-6) + 1[26-(-2)]$$

$$= 3(-40) - 2(-4) + 1(28)$$

$$= -120 + 8 + 28 = -84$$

Therefore,

$$x = \frac{D_x}{D} = \frac{28}{28} = 1, \quad y = \frac{D_y}{D} = -\frac{56}{28} = -2, \quad z = \frac{D_z}{D} = -\frac{84}{28} = -3$$

The solution is $x = 1, y = -2, z = -3$; this is also written as $(1, -2, -3)$.

If the denominator determinant, *D*, has a value of zero, then the system is either inconsistent or dependent. The system is dependent if all the determinants have a value of zero. The system is inconsistent if at least one of the determinants, D_x, D_y, or D_z, has a value not equal to zero and the denominator determinant has a value of zero.

Translating Word Problems into Three-Variable Systems

To understand the reasoning behind solving three-variable systems, Common Core Mathematics presents real-life scenarios where such systems can be created. Carefully read what the question is asking before solving the equation. Set up a three-variable system for each problem below and solve that system by choosing an appropriate method.

Example 1: A change machine in a laundromat contains a total of 24 coins in nickels, dimes, and quarters. There are three times as many dimes as there are nickels and quarters combined. The total value of all of the coins is \$2.90. Find the quantity of each type of coin in the change machine.

The problem translates into the following system, where d is the number of dimes, n is the number of nickels, and q is the number of quarters.

Solve this system using elimination. Write all equations in standard form: $ax + by + cz = d$.

$$\begin{cases} d+n+q=24 & (1) \\ 0.10d+0.05n+0.25q=2.90 & (2) \\ d-3n-3q=0 & (3) \end{cases}$$

Use equations (1) and (3) to eliminate d.

$$\begin{aligned} d+n+q&=24 \quad \text{Multiply by} -1 \\ d-3n-3q&=0 \end{aligned}$$

$$\begin{aligned} -d-n-q&=-24 \\ d-3n-3q&=0 \\ \hline -4n-4q&=-24 \quad (4) \end{aligned}$$

Use equations (2) and (3) to eliminate d.

$$\begin{aligned} 0.10d+0.05n+0.25q&=2.90 \quad \text{Multiply by } 100 \\ d-3n-3q&=0 \quad \text{Multiply by} -10 \end{aligned}$$

$$\begin{aligned} 10d+5n+25q&=290 \\ -10d+30n+30q&=0 \\ \hline 35n+55q&=290 \quad (5) \end{aligned}$$

Use equations (4) and (5) to eliminate n.

$$-4n - 4q = -24 \quad \text{Multiply by 35}$$
$$35n + 55q = 290 \quad \text{Multiply by 4}$$

$$\begin{aligned} -140n - 140q &= -840 \\ \underline{140n + 220q} &\underline{= 1160} \\ 80q &= 320 \\ q &= 4 \end{aligned}$$

Substitute $q = 4$ into equation (4).

$$\begin{aligned} -4n - 4q &= -24 \\ -4n - 4(4) &= -24 \\ -4n - 16 &= -24 \\ -4n &= -8 \\ n &= 2 \end{aligned}$$

Substitute $n = 2$ and $q = 4$ into equation (1).

$$\begin{aligned} d + n + q &= 24 \\ d + 2 + 4 &= 24 \\ d + 6 &= 24 \\ d &= 18 \end{aligned}$$

There are 18 dimes, 2 nickels, and 4 quarters in the change machine.

Example 2: A rectangular box has a base whose perimeter is 30 inches. The perimeter of the front face is 24 inches and the perimeter of a side wall is 26 inches. Find the dimensions of the box.

The dimensions of the base are the length and the width of the box. The dimensions of the front face are the length and the height of the box. The dimensions of a side wall are the width and the height of the box.

Solve this system using Cramer's Rule. Using l, w, and h, respectively, for the length, width, and height, the system becomes

$$\begin{aligned} 2l+2w&=30 \quad (1)\\ 2l+2h&=24 \quad (2)\\ 2w+2h&=26 \quad (3) \end{aligned}$$

$$D=\begin{vmatrix} 2 & 2 & 0 \\ 2 & 0 & 2 \\ 0 & 2 & 2 \end{vmatrix}=2\begin{vmatrix} 0 & 2 \\ 2 & 2 \end{vmatrix}-2\begin{vmatrix} 2 & 0 \\ 2 & 2 \end{vmatrix}+0\begin{vmatrix} 2 & 0 \\ 0 & 2 \end{vmatrix}$$

$$=2(0-4)-2(4-0)+0(4-0)$$

$$=-8-8+\ 0=-16$$

$$D_l=\begin{vmatrix} 30 & 2 & 0 \\ 24 & 0 & 2 \\ 26 & 2 & 2 \end{vmatrix}=30\begin{vmatrix} 0 & 2 \\ 2 & 2 \end{vmatrix}-24\begin{vmatrix} 2 & 0 \\ 2 & 2 \end{vmatrix}+26\begin{vmatrix} 2 & 0 \\ 0 & 2 \end{vmatrix}$$

$$=\ 30(0-4)-24(4-0)+26(4-0)$$

$$=-120-96+104=-112$$

$$D_w=\begin{vmatrix} 2 & 30 & 0 \\ 2 & 24 & 2 \\ 0 & 26 & 2 \end{vmatrix}=2\begin{vmatrix} 24 & 2 \\ 26 & 2 \end{vmatrix}-2\begin{vmatrix} 30 & 0 \\ 26 & 2 \end{vmatrix}+0\begin{vmatrix} 30 & 0 \\ 24 & 2 \end{vmatrix}$$

$$=\ 2(48-52)-2(60-0)+0(60-0)$$

$$=-8-120+0=-128$$

$$D_h=\begin{vmatrix} 2 & 2 & 30 \\ 2 & 0 & 24 \\ 0 & 2 & 26 \end{vmatrix}=2\begin{vmatrix} 0 & 24 \\ 2 & 26 \end{vmatrix}-2\begin{vmatrix} 2 & 30 \\ 2 & 26 \end{vmatrix}+0\begin{vmatrix} 2 & 30 \\ 0 & 24 \end{vmatrix}$$

$$=\ 2(0-48)-2(52-60)+0(48-0)$$

$$=-96+16+0=-80$$

$$\frac{D_l}{D}=\frac{-112}{-16}=7 \qquad \frac{D_w}{D}=\frac{-128}{-16}=8 \qquad \frac{D_h}{D}=\frac{-80}{-16}=5$$

The box has a length of 7 inches, a width of 8 inches, and a height of 5 inches.

Example 3: A triangle has angles x, y, and z. The measure of angle y is 10 degrees smaller than the measure of three times angle x. The measure of angle y is half the measure of angle z. Find the measure of all three angles in the triangle.

The sum of the measures of the angles in any triangle is 180°. If the measure of angle y is half the measure of angle z, then the measure of angle z is twice the measure of angle y. Solve this system by substitution. The problem translates into the following system:

$$\begin{aligned} x+y+z &= 180 \qquad (1) \\ y &= 3x-10 \quad (2) \\ z &= 2y \qquad (3) \end{aligned}$$

If $z = 2y$ and $y = 3x - 10$, then $z = 2(3x - 10)$, or $z = 6x - 20$.

Substitute $z = 6x - 20$ and $y = 3x - 10$ into equation (1).

$$\begin{aligned} x+y+z &= 180 \\ x+3x-10+6x-20 &= 180 \\ 10x-30 &= 180 \\ 10x &= 210 \\ x &= 21 \end{aligned}$$

Substitute $x = 21$ into equation (2) to get

$$\begin{aligned} y &= 3x-10 \\ y &= 3(21)-10 = 63-10 = 53 \end{aligned}$$

Substitute $y = 53$ into equation (3) to get

$$\begin{aligned} z &= 2y \\ z &= 2(53) = 106 \end{aligned}$$

The measures of the angles are 21°, 53°, and 106°, respectively.

Chapter Check-Out

Questions

1. Solve this system of equations by the method of elimination.

$$\begin{cases} 5x-2y+3z=22 \\ 2x+3y-4z=-2 \\ 3x+4y+z=30 \end{cases}$$

2. Solve this system of equations by the method of elimination.

$$\begin{cases} 5x+3y-7z=0 \\ 7x+5y-3z=16 \\ 3x-5y+2z=-8 \end{cases}$$

3. Use Cramer's Rule to solve this system of equations.

$$\begin{cases} 2x+3y-z=-7 \\ x-2y+z=-2 \\ 3x+y+2z=-7 \end{cases}$$

4. Use matrices to solve this system of equations. (***Note:*** When a variable term is missing, the coefficient of that variable is zero.)

$$\begin{cases} 2x-3z=3 \\ x-2y=-4 \\ 3y-2z=9 \end{cases}$$

Answers

1. (3, 4, 5)
2. (1, 3, 2)
3. (−3, 0, 1)
4. (6, 5, 3)

Chapter 5

POLYNOMIAL ARITHMETIC

Chapter Check-In

- ❑ Adding and subtracting polynomials
- ❑ Multiplying polynomials
- ❑ Working with special products of binomials
- ❑ Dividing polynomials
- ❑ Using synthetic division

Common Core Standard: Seeing the Structure in Expressions

Interpret the structure of expressions (A.SSE). Perform arithmetic operations on polynomials. Understand the relationship between zeros and factors of polynomials (A.APR).

Common Core Mathematics suggests an understanding of the "arithmetic" of polynomials and shows structural similarities between the system of polynomials and the system of integers. The processes are similar by using variables in addition to numerals. This understanding is crucial to your ability to solve equations involving variables with exponents presented in subsequent chapters.

Polynomials are expressions containing one or more than one term, with each term separated from the preceding term by a plus or a minus sign. The exponents on the variables in a polynomial are always whole numbers. A polynomial has no maximum length. Some arithmetic operations with polynomials use common sense, but others require special techniques. All are presented in this chapter.

Adding and Subtracting Polynomials

In order to add and subtract polynomials successfully, you must understand what monomials, binomials, and trinomials are; what constitutes "like terms"; and the difference between ascending and descending order.

Monomials, binomials, and trinomials

A **monomial** is an expression that could be a numeral, a variable, or the product of numerals and variables. If the expression has variables, certain restrictions apply to make it a monomial.

- Variables must have whole-number exponents.
- Variables do not appear under a radical sign in simplified radical expressions.
- Denominators do not contain variables.

The following expressions are examples of monomials.

$$-12,\ a,\ 3t^2,\ \frac{5}{8}x^2y^3,\ y^3,\ \frac{2(x+y^2)}{3}$$

The following are expressions that are *not* monomials.

x^{-2} (The variable has an exponent that is not a whole number.)

$\sqrt{x}$ (A variable is under a simplified radical.)

$\dfrac{3(x+2)}{x+3}$ (The denominator has a variable.)

A **binomial** is an expression that is the sum of two monomials.

A **trinomial** is an expression that is the sum of three monomials.

A **polynomial** is an expression that is a monomial or the sum of two or more monomials.

Like terms or similar terms

Two or more monomials with identical variable expressions are called *like terms* or *similar terms.* The following are like terms, since their variable expressions are all x^2y:

$$5x^2y, -3x^2y, \frac{2}{7}x^2y$$

The following are not like terms, since their variable expressions are not all the same:

$$-5x^2y^2, 4x^2y, \frac{1}{2}xy^2$$

In order to add monomials, they must be like terms. *Unlike terms cannot be added together.* To add like terms, follow these steps.

1. Add their numerical coefficients (number preceding the letter).
2. Keep the variable expression.

Example 1: Find the following sums.

(a) $4x^2y + 8x^2y = 12x^2y$

(b) $-9abc + 3abc = -6abc$

(c) $9xy + 7x - 28xy - 4x = -19xy + 3x$

Note that in item (c), because $-19xy$ and $3x$ are *unlike* terms, they *cannot* be added together.

Ascending and descending order

When working with polynomials that involve only one variable, the general practice is to write them so that the exponents on the variable decrease from left to right. The polynomial is then said to be written in **descending order.**

When a polynomial in one variable is written so that the exponents increase from left to right, it is referred to as being written in **ascending order.**

Example 2: Rewrite the following polynomial in descending powers of x.

$$4y^4 + 12 - 15x^2 + 13x^3y + 17xy^2$$

This becomes

$$13x^3y - 15x^2 + 17xy^2 + 4y^4 + 12$$

To add two or more polynomials, add like terms and arrange the answer in descending (or ascending, if specified) powers of one variable.

Example 3: Find the sum:

$$\begin{aligned}&(x^2+x^3-3x)+(4-5x^2+3x^3)+(10-8x^2-5x)\\&=(x^3+3x^3)+(x^2-5x^2-8x^2)+(-3x-5x)+(4+10)\\&=4x^3-12x^2-8x+14\end{aligned}$$

This problem can also be added vertically. First, rewrite each polynomial in descending order, one above the other, placing like terms in the same column.

$$\begin{array}{r}x^3+\ \ x^2-3x\\3x^3-5x^2+4\\-8x^2-5x+10\\\hline 4x^3-12x^2-8x+14\end{array}$$

To subtract one polynomial from another, add its opposite.

Example 4: Subtract $(4x^2 - 7x + 3)$ from $(6x^2 + 4x - 9)$.

Worked horizontally,

$$\begin{aligned}(6x^2+4x-9)-(4x^2-7x+3)&=6x^2+4x-9-4x^2+7x-3\\&=2x^2+11x-12\end{aligned}$$

Worked vertically,

$$\begin{array}{rr}6x^2+4x-9\rightarrow & 6x^2+4x-9\\-(4x^2-7x+3)\rightarrow & \underline{-4x^2+7x-3}\\ & 2x^2+11x-12\end{array}$$

Multiplying Polynomials

The following are rules regarding the multiplying of variable expressions. These rules are referred to as the *laws of exponents*. Exponents are also referred to as *powers*.

The Laws of Exponents—Multiplication

Rule 1: To multiply monomials with the same base, keep the base and add the exponents.

$$(x^a)(x^b) = x^{a+b}$$

Rule 2: To raise a base to a power, keep the base and multiply the exponents.

$$(x^a)^b = x^{ab}$$

Rule 3: To raise a product to a power, raise each factor in the product to that power.

$$(xy)^a = x^a y^a$$

Example 1: Simplify each of the following multiplication problems and state which of the preceding rules were applied.

(a) $yy^5 = y^{1+5} = y^6$ (Rule 1)

(b) $(x^4)^3 = x^{4 \cdot 3} = x^{12}$ (Rule 2)

(c) $(-2x^4y^2z^3)^5 = (-2)^5x^{4 \cdot 5}y^{2 \cdot 5}z^{3 \cdot 5} = -32x^{20}y^{10}z^{15}$ (Rule 3)

(d) $a^3(a^2b^3)^4 = a^3(a^2)^4(b^3)^4$ (Rule 3)

$= a^3a^8b^{12}$ (Rule 2)

$= a^{11}b^{12}$ (Rule 1)

To multiply monomials together, follow these steps:

1. Multiply the numerical coefficients together.
2. Multiply the variables together, using the *laws of exponents.*
3. Write the results as a product.

Example 2: Simplify each of the following.

(a) $(4x^2)(3x^3) = (4 \cdot 3)(x^2 \cdot x^3) = 12x^5$

(b) $(-8a^3b^2)(2a^2b^2)^3 = (-8a^3b^2)(8a^6b^6) = -64a^9b^8$

To multiply polynomials together, multiply each term in one polynomial by each term in the other polynomial; simplify if possible.

Example 3: Multiply each of the following (use the distributive property).

(a) $5x(3x^2 - 4x + 2)$

Expression (a) worked horizontally:

$$5x(3x^2 - 4x + 2) = 15x^3 - 20x^2 + 10x$$

Expression (a) worked vertically:

$$\begin{array}{lr} & 3x^2 - 4x + 2 \\ \times & 5x \\ \hline & 15x^3 - 20x^2 + 10x \end{array}$$

(b) $(4x - 2)(3x + 5)$

Expression (b) worked horizontally:

$$\begin{aligned} (4x - 2)(3x + 5) &= 12x^2 + 20x - 6x - 10 \\ &= 12x^2 + 14x - 10 \end{aligned}$$

Expression (b) worked vertically:

$$\begin{array}{lr} & 3x + 5 \\ \times & 4x - 2 \\ \hline & -6x - 10 \\ & 12x^2 + 20x \\ \hline & 12x^2 + 14x - 10 \end{array}$$

(c) $(x + y)(x^2 - xy + y^2)$

Expression (c) worked horizontally:

$$(x+y)(x^2-xy+y^2)=x^3-x^2y+xy^2+x^2y-xy^2+y^3$$
$$=x^3+y^3$$

Expression (c) worked vertically:

$$\begin{array}{r} x^2-xy+y^2 \\ \times \qquad x+y \\ \hline x^2y-xy^2+y^3 \\ x^3-x^2y+xy^2 \qquad \\ \hline x^3 \qquad\qquad\qquad +y^3 \end{array}$$

Working with Special Products of Binomials

It is often more efficient to memorize and recognize patterns of special products of binomials. Two binomials with the same first terms but opposite second terms are called **conjugates** of each other. Following are examples of conjugates:

$$3x+2 \quad \text{and} \quad 3x-2$$
$$-5a-4b \quad \text{and} \quad -5a+4b$$

Example 1: Find the product of the following conjugates.

(a) $(3x+2)(3x-2)=9x^2-6x+6x-4$
$=9x^2-4$

(b) $(-5a-4b)(-5a+4b)=25a^2-20ab+20ab-16b^2$
$=25a^2-16b^2$

Notice that when conjugates are multiplied together, the answer is the difference of the squares of the terms in the original binomials.

The product of conjugates produces a special pattern referred to as a **difference of squares.** In general,

$$(x + y)(x - y) = x^2 - y^2$$

The squaring of a binomial also produces a special pattern.

Example 2: Simplify each of the following.

(a) $(4x+3)^2 = (4x+3)(4x+3)$

$$= 16x^2 + 12x + 12x + 9$$

$$= 16x^2 + 2(12x) + 9$$

$$= 16x^2 + 24x + 9$$

(b) $(6a-7b)^2 = (6a-7b)(6a-7b)$

$$= 36a^2 - 42ab - 42ab + 49b^2$$

$$= 36a^2 - 2(42ab) + 49b^2$$

$$= 36a^2 - 84ab + 49b^2$$

First, notice that the answers are trinomials. Second, notice that there is a pattern in the terms:

- The first and last terms are the squares of the first and last terms of the binomial.
- The middle term is *twice* the product of the two terms in the binomial.

The pattern produced by squaring a binomial is referred to as a **square trinomial.** In general,

$$(x+y)^2 = x^2 + 2xy + y^2$$

and

$$(x-y)^2 = x^2 - 2xy + y^2$$

Example 3: Simplify the following special binomial products.

(a) $(3x + 4y)^2 = 9x^2 + 24xy + 16y^2$

(b) $(6x + 11)(6x - 11) = 36x^2 - 121$

Dividing Polynomials

There are also laws of exponents for division.

> ### The Laws of Exponents—Division
>
> **Rule 4:** $\left(\frac{a}{b}\right)^n = \frac{a^n}{b^n}$
>
> **Rule 5:** To divide monomials with the same base, keep the base and subtract the exponents.
>
> $$\frac{a^m}{a^n} = a^{m-n} \text{ as long as } a \neq 0$$
>
> **Rule 6:** Monomials that have negative exponents become their reciprocal with a positive exponent.
>
> $$a^{-n} = \frac{1}{a^n} \text{ and } \frac{1}{a^{-n}} = a^n \text{ for all } (a \neq 0)$$

Example 1: Simplify each of the following division expressions and find the pattern involving the exponents.

(a) $\frac{a^8}{a^2} = \frac{\overset{1}{\cancel{a}}\,\overset{1}{\cancel{a}}\,aaaaaa}{\underset{1}{\cancel{a}}\,\underset{1}{\cancel{a}}} = a^6$

(b) $\frac{a^8}{a^3} = \frac{\overset{1}{\cancel{a}}\,\overset{1}{\cancel{a}}\,\overset{1}{\cancel{a}}\,aaaaa}{\underset{1}{\cancel{a}}\,\underset{1}{\cancel{a}}\,\underset{1}{\cancel{a}}} = a^5$

Example 2 illustrates Rule 5.

Example 2: Simplify each of the following:

(a) $\frac{a^5}{a^5} = a^{5-5} = a^0 = 1$

(b) $\frac{a^5}{a^6} = a^{5-6} = a^{-1}$, also $\frac{a^5}{a^6} = \frac{\overset{1}{\cancel{a}}\,\overset{1}{\cancel{a}}\,\overset{1}{\cancel{a}}\,\overset{1}{\cancel{a}}\,\overset{1}{\cancel{a}}}{\underset{1}{\cancel{a}}\,\underset{1}{\cancel{a}}\,\underset{1}{\cancel{a}}\,\underset{1}{\cancel{a}}\,\underset{1}{\cancel{a}}\,a} = \frac{1}{a}$, so $a^{-1} = \frac{1}{a}$

(c) $\frac{a^5}{a^8} = a^{5-8} = a^{-3}$, also $\frac{a^5}{a^8} = \frac{\overset{1}{\cancel{a}}\,\overset{1}{\cancel{a}}\,\overset{1}{\cancel{a}}\,\overset{1}{\cancel{a}}\,\overset{1}{\cancel{a}}}{\underset{1}{\cancel{a}}\,\underset{1}{\cancel{a}}\,\underset{1}{\cancel{a}}\,\underset{1}{\cancel{a}}\,\underset{1}{\cancel{a}}\,aaa} = \frac{1}{a^3}$, so $a^{-3} = \frac{1}{a^3}$

Generally, when simplifying expressions, write the final result without the use of negative exponents. Example 3 illustrates Rule 6.

Example 3: Simplify each of the following.

(a) $\dfrac{x^{2a}}{x^{3a}} = x^{2a-3a} = x^{-a} = \dfrac{1}{x^a}$

(b) $\dfrac{3^{4x-3}}{3^{2x-4}} = 3^{(4x-3)-(2x-4)} = 3^{4x-3-2x+4} = 3^{2x+1}$

To divide a monomial by another monomial, follow this procedure:

1. Divide the numerical coefficients.
2. Divide the variables.
3. Write the results as a quotient.

Example 4: Simplify the following expressions by dividing correctly.

(a) $\dfrac{12x^8}{4x^3} = \dfrac{12}{4} \cdot \dfrac{x^8}{x^3} = 3x^5$

(b) $\dfrac{-27w^3t^7}{-3w^3t^{12}} = \dfrac{-27}{-3} \cdot \dfrac{w^3}{w^3} \cdot \dfrac{t^7}{t^{12}} = 9 \cdot 1 \cdot \dfrac{1}{t^5} = \dfrac{9}{t^5}$

(c) $\dfrac{-15x^3y^4}{10y^5z} = \dfrac{-15}{10} \cdot \dfrac{x^3}{1} \cdot \dfrac{y^4}{y^5} \cdot \dfrac{1}{z} = \dfrac{-3x^3}{2yz}$

To divide a polynomial by a monomial, divide each term of the polynomial by the monomial (applying the laws of exponents).

Example 5: Simplify the given expression by dividing correctly.

$$\frac{15r^2s - 7rs^2 + 6s^3}{-3r^2} = \frac{15r^2s}{-3r^2} - \frac{7rs^2}{-3r^2} + \frac{6s^3}{-3r^2}$$

$$= -5s + \frac{7s^2}{3r} - \frac{2s^3}{r^2}$$

To divide a polynomial by a polynomial, a procedure similar to long division in arithmetic is used. The procedure calls for four steps: divide, multiply, subtract, and bring down. This procedure is repeated until there is no value left to bring down.

Example 6: Divide $x^2 + 3x^3 - 5$ by $4 + x$.

First, arrange both polynomials in descending order, filling in a placeholder for any missing terms.

$$x+4\overline{)3x^3+x^2+0x-5}$$

$\uparrow$ placeholder, since there is no "x" term

Now *divide* $3x^3$ by x and bring this partial quotient to the top as the first part of the answer.

$$\frac{3x^3}{x}=3x^2 \longrightarrow \begin{array}{r} 3x^2 \\ x+4\overline{)3x^3+x^2+0x-5} \end{array}$$

Multiply $3x^2$ by the divisor $x + 4$ and place these in the columns with like terms.

$$3x^2(x+4)=3x^3+12x^2 \rightarrow \begin{array}{r} 3x^2 \\ x+4\overline{)3x^3+\quad x^2+0x-5} \\ \underline{3x^3+12x^2} \end{array}$$

Subtract. Remember, to subtract is to add the opposite.

$$\text{subtract} \rightarrow \begin{array}{r} 3x^2 \\ x+4\overline{)3x^3+\quad x^2+0x-5} \\ \underline{-(3x^3+12x^2)} \\ -11x^2 \end{array}$$

Bring down the next term and start the procedure again.

$$\begin{array}{rl}
 & 3x^2 \\
x+4 & \overline{)3x^3 + \quad x^2 + 0x - 5} \\
 & \underline{-(3x^3 + 12x^2)} \;\downarrow \\
 & \quad -11x^2 + 0x
\end{array}$$

Divide. $\frac{-11x^2}{x} = -11x$

$$\begin{array}{rl}
 & 3x^2 - 11x \\
x+4 & \overline{)3x^3 + \quad x^2 + \quad 0x - 5} \\
 & \underline{-\left(3x^3 + 12x^2\right)} \\
 & \quad -11x^2 + \quad 0x \\
 & \quad \underline{-\left(-11x^2 - 44x\right)} \;\downarrow \\
 & \qquad\qquad 44x - 5
\end{array}$$

Multiply and subtract.

$-11x(x+4) = -11x^2 - 44x \longrightarrow -\left(-11x^2 - 44x\right)$

Bring down.

Since a value was brought down, start the process over again.

Divide. $\frac{44x}{x} = 44$

$$\begin{array}{rl}
 & 3x^2 - 11x \qquad + 44 \\
x+4 & \overline{)3x^3 + \quad x^2 + \quad 0x - 5} \\
 & \underline{-\left(3x^3 + 12x^2\right)} \\
 & \quad -11x^2 + \quad 0x \\
 & \quad \underline{-\left(-11x^2 - 44x\right)} \;\downarrow \\
 & \qquad\qquad 44x - 5 \\
 & \qquad\qquad \underline{-\left(44x + 176\right)} \\
 & \qquad\qquad\qquad -181
\end{array}$$

Multiply and subtract.

$44\left(x+4\right) = 44x + 176 \longrightarrow -\left(44x + 176\right)$

At this point, there is no term to bring down. The -181 is the remainder.

This process is identical to long division with numerals, and as in arithmetic, remainders are written over the divisor. So the final answer to the division problem is

$$3x^2 - 11x + 44 - \frac{181}{x+4}$$

Example 7: Divide $64x^3 - 27$ by $4x - 3$.

First, divide $64x^3$ by $4x$.

$$\frac{64x^3}{4x} = 16x^2$$

Write it above the first term under the long-division symbol. Then multiply and subtract.

$$16x^2(4x-3) = 64x^3 - 48x^2$$

Next, divide $48x^2$ by $4x$

$$\left(\text{think: } \frac{48x^2}{4x} = 12x\right).$$

$$\begin{array}{r|l}
 & 16x^2 + 12x + 9 \\ \hline
4x - 3 & 64x^3 + 0x^2 + 0x - 27 \\
 & -(64x^3 - 48x^2) \\ \hline
 & \quad 48x^2 + 0x \\
 & \quad -(48x^2 - 36x) \\ \hline
 & \qquad 36x - 27 \\
 & \qquad -(36x - 27) \\ \hline
 & \qquad\qquad 0
\end{array}$$

Write that above the next term under the long-division symbol, with a plus sign separating it from the first one. Then multiply and subtract.

Finally, repeat the sequence with the $36x$.

$$\frac{36x}{4x} = 9$$

$$9(4x-3) = 36x - 27$$

The answer (**quotient**) is $16x^2 + 12x + 9$. Notice that there is no remainder.

Using Synthetic Division

Synthetic division is a shortcut for polynomial division when the divisor is of the form $x - a$. Only numeric coefficients of the dividend are used when dividing with synthetic division.

Example 1: Divide $(2x - 11 + 3x^3)$ by $(x - 3)$.

First, this problem is done in the traditional manner. Then it is done by using the synthetic division method.

In the traditional manner,

$$\begin{array}{r|l} & 3x^2 + 9x + 29 \\ \hline x-3 & 3x^3 + 0x^2 + 2x - 11 \\ & \underline{-(3x^3 - 9x^2)} \downarrow \\ & \quad 9x^2 + 2x \\ & \quad \underline{-(9x^2 - 27x)} \downarrow \\ & \qquad 29x - 11 \\ & \qquad \underline{-(29x - 87)} \\ & \qquad\qquad 76 \end{array}$$

The answer is $3x^2 + 9x + 29 + \dfrac{76}{x-3}$.

To do the problem using synthetic division, follow these steps:

1. Write the polynomial being divided in descending order. Then write only its coefficients and constant, using 0 for any missing terms.

$$3x^3 + 0x^2 + 2x - 11$$

$$3 \quad 0 \quad 2 \quad -11$$

2. Write the constant, a, of the divisor, $x - a$, to the left. In this problem, $a = 3$ because you use the additive inverse of the constant. (Remember, the additive inverse of -3 is 3.)

$$\underline{3|} \quad 3 \quad 0 \quad 2 \quad -11$$

3. Bring down the first coefficient as shown.

$$\begin{array}{r|rrrr} 3 & 3 & 0 & 2 & -11 \\ & \downarrow & & & \\ \hline & 3 & & & \end{array}$$

4. Multiply the first coefficient by the divisor, 3. Then write this product under the second coefficient.

$$(3\times 3=9)\longrightarrow \begin{array}{r|rrrr} 3 & 3 & 0 & 2 & -11 \\ & & 9 & & \\ \hline & 3 & & & \end{array}$$

5. Add the second coefficient with the product and write the sum as shown.

$$\begin{array}{r|rrrr} 3 & 3 & 0 & 2 & -11 \\ & & 9 & & \\ \hline & 3 & 9 & & \end{array}$$

6. Continue this process of multiplying and adding until there is a sum for the last column.

$$\begin{array}{r|rrrr} 3 & 3 & 0 & 2 & -11 \\ & & 9 & 27 & 87 \\ \hline & 3 & 9 & 29 & 76 \end{array}$$

The numbers along the bottom row are the coefficients of the quotient with the powers of x in descending order. The last coefficient is the remainder. The first power is one less than the highest power of the polynomial that was being divided.

The division answer is $3x^2+9x+29+\dfrac{76}{x-3}$.

Example 2: Divide $(5x^4 + 6x^3 - 9x^2 - 7x + 6)$ by $(x + 2)$ using synthetic division.

To put the divisor, $x + 2$, into the form $x - a$, use the constant's opposite. Therefore, $x + 2$ becomes $x - (-2)$.

$$\begin{array}{r|rrrrr} -2 & 5 & 6 & -9 & -7 & 6 \\ & & -10 & 8 & 2 & 10 \\ \hline & 5 & -4 & -1 & -5 & 16 \end{array}$$

The answer is $5x^3 - 4x^2 - x - 5 + \dfrac{16}{x+2}$.

Related to the concept of synthetic division is **synthetic substitution**. This process is a shortcut for evaluating a polynomial given a value of the variable.

Example 3: Evaluate $x^3 - 3x^2 + 2x - 9$ when $x = -4$.

$$(-4)^3 - 3(-4)^2 + 2(-4) - 9$$
$$-64 - 3(16) + 2(-4) - 9$$
$$-64 - 48 - 8 - 9$$
$$-129$$

Synthetic substitution allows us to use the same setup as synthetic division, using the value of the variable in the box on the left. The final value at the far right of the answer is the value of the polynomial:

$$\begin{array}{r|rrrr} -4 & 1 & -3 & 2 & -9 \\ & & -4 & 28 & -120 \\ \hline & 1 & -7 & 30 & -129 \end{array}$$

Chapter Check-Out

Questions

1. Find the following difference:
   $(5x^4 + 3x^3 + 2x^2 - 4x) - (2x^4 + 5x^2 + 4x - 6)$

2. Find the following sum:
   $(5ab^2 - 2ab + 3a^2b) + (3ab - 2ab^2 + a^2b)$

3. Simplify each of the following multiplication problems:
   (a) $b^2 \cdot b^3$
   (b) $(y^4)^3$
   (c) $(-3x^4y^3z^2)^3$
   (d) $(2x - 3)(x + 4)$
   (e) $(3x + 2)(3x - 2)$

4. Explain the difference between synthetic division and synthetic substitution.

5. What is the advantage of using synthetic division over using long division?

6. When using synthetic division, how can you tell if the divisor is a factor of the dividend?

Answers

1. $3x^4 + 3x^3 - 3x^2 - 8x + 6$
2. $4a^2b + ab + 3ab^2$
3. **(a)** b^5; **(b)** y^{12}; **(c)** $-27x^{12}y^9z^6$; **(d)** $2x^2 + 5x - 12$; **(e)** $9x^2 - 4$
4. In synthetic division, the entire bottom row is your answer, whereas in synthetic substitution, only the final value is your answer.
5. Answers may vary. The advantage is that you only need to use and compute with the numerical coefficients.
6. If the final value is zero, then the divisor is a factor of the dividend because that is the value of the remainder. There is no remainder.

Chapter 6
FACTORING POLYNOMIALS

Chapter Check-In

- ❑ Factoring out the greatest common factor
- ❑ Factoring the difference of squares
- ❑ Factoring the differences of cubes and the sums of cubes
- ❑ Factoring various types of trinomials
- ❑ Factoring by grouping
- ❑ Solving equations by factoring

Common Core Standard: Seeing the Structure in Expressions

Arithmetic, Polynomials and Rational Expressions. Understand the relationship between zeros and factors of polynomials (A.APR.3). Understand solving equations as a process of reasoning and explain the reasoning (A.REI.1-2).

To factor a polynomial means to rewrite the polynomial as a product of simpler polynomials or of polynomials and monomials. Because polynomials may take many different forms, many different techniques are available for factoring them. In this chapter we explore all of the common methods of factoring polynomials. Then we see that factoring can be used as a powerful tool for solving equations of a degree higher than 1.

Greatest Common Factor

The first method of factoring is called factoring out the **GCF (greatest common factor)**.

Example 1: Factor $5x + 5y$.

Since each term in this polynomial involves a factor of 5, then $5x + 5y$ can be factored into $5(x + y)$. The answer can be verified by using the distributive property.

Example 2: Factor $24x^3 - 16x^2 + 8x$.

In this polynomial, x is a common factor for all three terms. Also, the numbers 24, -16, and 8 all have the common factors of 2, 4, and 8. The greatest common factor is, therefore, $8x$, which produces $8x(3x^2 - 2x + 1)$. This answer can be verified by using the distributive property.

Difference of Squares

Recall that the product of conjugates produces a pattern called a **difference of squares.**

$$(a + b) \text{ and } (a - b) \text{ are conjugates}$$

$$(a+b)(a-b) = \underbrace{a^2 - b^2}_{\substack{\text{difference}\\ \text{of squares}}}$$

Reversing that pattern allows us to factor the difference of squares back into its conjugates.

Example 1: Factor $x^2 - 16$.

This polynomial results from the subtraction of two values that are each the square of some expression.

$$x^2 = (x)^2 \text{ and } 16 = (4)^2$$

So

$$\underbrace{x^2 - 16}_{\substack{\text{difference}\\ \text{of squares}}} = (x)^2 - (4)^2$$

$$= \underbrace{(x+4)(x-4)}_{\text{conjugates}}$$

Note: Answers can be verified by the F.O.I.L. method.

Example 2: Factor $25x^2y^2 - 36z^2$.

$$25x^2y^2 = (5xy)^2 \text{ and } 36z^2 = (6z)^2$$

$$\begin{aligned}25x^2y^2 - 36z^2 &= (5xy)^2 - (6z)^2 \\ &= (5xy + 6z)(5xy - 6z)\end{aligned}$$

Example 3: Factor $(a + b)^2 - (c - d)^2$.

$$\begin{aligned}(a+b)^2 - (c-d)^2 &= [(a+b)+(c-d)][(a+b)-(c-d)] \\ &= (a+b+c-d)(a+b-c+d)\end{aligned}$$

Example 4: Factor $y^2 + 9$.

Note that $y^2 + 9$ is prime, or not factorable. Only the *difference* of two perfect squares is factorable.

Using more than one method of factoring

Many polynomials require more than one method of factoring to be completely factored into a product of polynomials. Because of this, a sequence of factoring methods must be used.

- First, try to factor by using the GCF.
- Second, try to factor by using the difference of squares.

Example 5: Factor $9x^2 - 36$.

$$9x^2 - 36 = \underbrace{9(x^2 - 4)}_{\text{difference of squares}} \qquad \text{GFC of 9}$$

$$= 9(x+2)(x-2)$$

Example 6: Factor $8(x + y)^2 - 18$.

$$8(x+y)^2 - 18 = 2[4(x+y)^2 - 9] \qquad \text{GFC of 2}$$

Note: $4(x + y)^2 = [2(x + y)]^2$ and $9 = 3^2$.

$$\begin{aligned}&= 2[2(x+y)+3][2(x+y)-3] \\ &= 2(2x+2y+3)(2x+2y-3)\end{aligned}$$

Differences of Cubes and Sums of Cubes

A polynomial in the form $a^3 + b^3$ is called a **sum of cubes.** A polynomial in the form $a^3 - b^3$ is called a **difference of cubes.** Both of these polynomials have similar factored patterns:

A sum of cubes:

$$a^3 + b^3 = (a + b)(a^2 - ab + b^2)$$

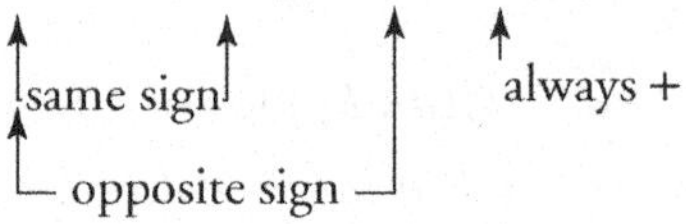

A difference of cubes:

$$a^3 - b^3 = (a - b)(a^2 + ab + b^2)$$

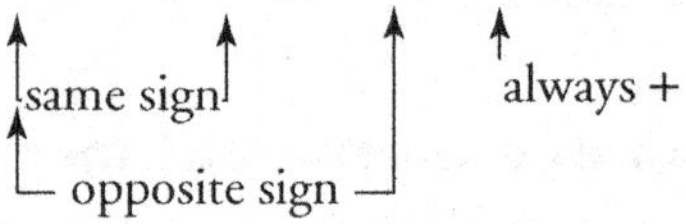

Example 1: Factor $x^3 + 125$.

$$\begin{aligned} x^3 + 125 &= (x)^3 + (5)^3 \\ &= (x+5)[x^2 - (x)(5) + 5^2] \\ &= (x+5)(x^2 - 5x + 25) \end{aligned}$$

Note: Answers can be verified by a double distribution of the first binomial over the second trinomial.

Example 2: Factor $8x^3 - 27$.

$$\begin{aligned} 8x^3 - 27 &= (2x)^3 - (3)^3 \\ &= (2x-3)[(2x)^2 + (2x)(3) + 3^2] \\ &= (2x-3)(4x^2 + 6x + 9) \end{aligned}$$

Example 3: Factor $2x^3 + 128y^3$.

First, find the GCF. The GCF = 2.

$$\begin{aligned} 2x^3 + 128y^3 &= 2(x^3 + 64y^3) \\ &= 2[(x)^3 + (4y)^3] \\ &= 2(x+4y)[x^2 - (x)(4y) + (4y)^2] \\ &= 2(x+4y)(x^2 - 4xy + 16y^2) \end{aligned}$$

Example 4: Factor $x^6 - y^6$.

First, notice that $x^6 - y^6$ is both a difference of squares and a difference of cubes.

$$x^6 - y^6 = (x^3)^2 - (y^3)^2 \qquad\qquad x^6 - y^6 = (x^2)^3 - (y^2)^3$$

In general, factor a difference of squares before factoring a difference of cubes.

$$\begin{aligned} x^6 - y^6 &= \underbrace{(x^3)^2 - (y^3)^2}_{\substack{\text{difference}\\ \text{of squares}}} \\ &= \underbrace{(x^3 + y^3)}_{\substack{\text{sum of}\\ \text{cubes}}}\underbrace{(x^3 - y^3)}_{\substack{\text{difference}\\ \text{of cubes}}} \\ &= [(x+y)(x^2 - xy + y^2)][(x-y)(x^2 + xy + y^2)] \\ &= (x+y)(x^2 - xy + y^2)(x-y)(x^2 + xy + y^2) \end{aligned}$$

Trinomials of the Form $x^2 + bx + c$

To factor polynomials of the form $x^2 + bx + c$, begin with two pairs of parentheses with x at the left of each.

$$(x \quad)(x \quad)$$

Next, find two integers whose product is c and whose sum is b and place them at the right of the parentheses.

Example 1: Factor $x^2 + 8x + 12$.

$$x^2 + 8x + 12 = (x \quad)(x \quad)$$

12 can be factored in a variety of ways:

$$(1)(12), (-1)(-12), (2)(6), (-2)(-6), (3)(4), (-3)(-4)$$

Only one of these pairs of factors adds to 8, namely (2)(6), so

$$x^2 + 8x + 12 = (x + 2)(x + 6)$$

Note: Answers can be verified by the F.O.I.L. method.

Example 2: Factor $x^2 - 7x - 18$.

-18 can be factored in the following ways:

$$(1)(-18), (-1)(18), (2)(-9), (-2)(9), (3)(-6), (-3)(6)$$

The only combination whose sum is also -7 is $(2)(-9)$, so

$$x^2 - 7x - 18 = (x + 2)(x - 9)$$

Example 3: Factor $x^2 - 6x + 9$.

9 can be factored as

$$(1)(9), (-1)(-9), (3)(3), (-3)(-3)$$

The only combination whose sum is -6 is $(-3)(-3)$, so

$$x^2 - 6x + 9 = (x - 3)(x - 3) = (x - 3)^2$$

Trinomials of the Form $ax^2 + bx + c$

Study this pattern for multiplying two binomials:

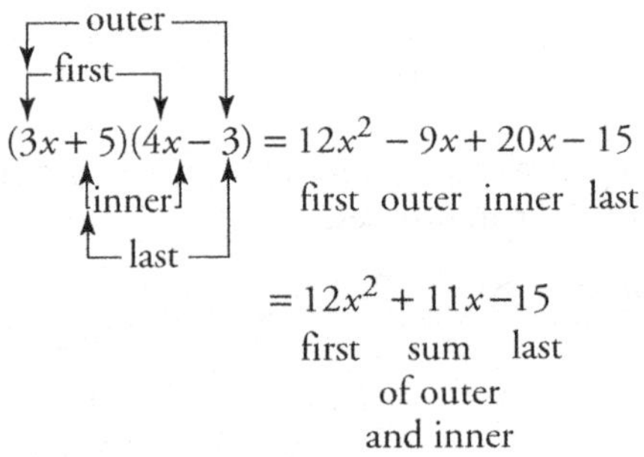

Example 1: Factor $2x^2 - 5x - 12$.

Begin by writing two pairs of parentheses.

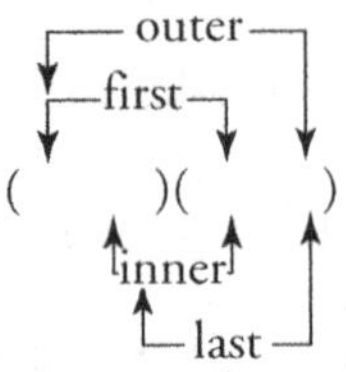

For the first positions, find two factors whose product is $2x^2$. For the last positions, find two factors whose product is -12. Following are the possibilities. The reason for the underlines will be explained shortly. With each possibility, the sum of outer and inner products is included.

1. $(x + 1)\underline{(2x - 12)}; -10x$
2. $(x - 1)\underline{(2x + 12)}; +10x$
3. $(x - 12)(2x + 1); -23x$
4. $(x + 12)(2x - 1); +23x$
5. $(x + 2)\underline{(2x - 6)}; -2x$
6. $(x - 2)\underline{(2x + 6)}; +2x$
7. $(x - 6)\underline{(2x + 2)}; -10x$
8. $(x + 6)\underline{(2x - 2)}; +10x$
9. $(x + 3)\underline{(2x - 4)}; +2x$
10. $(x - 3)\underline{(2x + 4)}; -2x$
11. $(x - 4)(2x + 3); -5x$
12. $(x + 4)(2x - 3); +5x$

Only possibility 11 will multiply out to produce the original polynomial. Therefore,

$$2x^2 - 5x - 12 = (x - 4)(2x + 3)$$

Because many possibilities exist, some shortcuts are advisable:

- **Shortcut 1:** Be sure the GCF, if there is one, has been factored out.
- **Shortcut 2:** Try factors closest to each other first. For example, when considering factors of 12, try 3 and 4 before trying 6 and 2 and try 6 and 2 before trying 1 and 12.
- **Shortcut 3:** Avoid creating binomials that will have a GCF within them. This shortcut eliminates possibilities 1, 2, 5, 6, 7, 8, 9, and 10 (look at the underlined binomials; their terms each have some common factor), leaving only four possibilities to consider. Of the four remaining possibilities, 11 and 12 would be considered first using shortcut 2.

Example 2: Factor $8x^2 - 26x + 20$.

$$8x^2 - 26x + 20 = 2(4x^2 - 13x + 10) \qquad \text{GCF of 2}$$

For first factors, begin with $2x$ and $2x$ (closest factors). For last factors, begin with -5 and -2 (closest factors and the product is positive; since the middle term is negative, both factors need to be negative).

$$(2x - 5)(2x - 2)$$

Shortcut 3 eliminates this possibility.

Now, try -1 and -10 for last factors.

$$(2x - 1)(2x - 10)$$

Shortcut 3 eliminates this possibility.

Now, try $1x$ and $4x$ for first factors and go back to -5 and -2 as last factors.

$$(x - 5)(4x - 2)$$

Shortcut 3 eliminates this possibility. But because x and $4x$ are different factors, switching the -5 and -2 produces different results, as shown in the following:

outer $= -5x$

$$(x - 2)(4x - 5) \quad \text{sum of outer and inner} = -13x$$

inner $= -8x$

Therefore, $8x^2 - 26x + 20 = 2(x - 2)(4x - 5)$.

Square Trinomials

Recall that

$$\underbrace{(x+y)^2}_{\text{the square of a binomial}} = \underbrace{(x)^2 + 2(x)(y) + (y)^2}_{\text{square trinomial}}$$

Therefore, if a trinomial is of the form $(x)^2 + 2(x)(y) + (y)^2$, it can be factored into the square of a binomial.

Example 1: Is $4x^2 - 20x + 25$ a square trinomial? If so, factor it into the square of some binomial.

$$4x^2 = (2x)^2 \text{ and } 25 = (-5)^2 \text{ and } -20x = 2(2x)(-5)$$

So, it is a square trinomial, which factors as follows.

$$4x^2 - 20x + 25 = (2x - 5)^2$$

Example 2: Is $x^2 + 10x + 9$ a square trinomial?

$$x^2 = (x)^2 \text{ and } 9 = 3^2 \text{ but } 10x \neq 2(x)(3)$$

So, it is not a square trinomial. But $x^2 + 10x + 9$ is factorable.

$$x^2 + 10x + 9 = (x + 1)(x + 9)$$

Factoring by Grouping

To attempt to factor a polynomial of four or more terms with no common factor, first rewrite it in groups. Each group may possibly be separately factored, and the resulting expression may possibly lend itself to further factorization if a GCF or special form is created.

Example 1: Factor $ay + az + by + bz$.

This polynomial has four terms with no common factor. It could be put into either two groups of two terms or two groups with three terms in one group and one term in the other group. One such arrangement is

$$(ay+az)+(by+bz)$$

$$ay+az=a(y+z) \quad \text{GCF of } a$$

$$by+bz=b(y+z) \quad \text{GCF of } b$$

The two new terms have a GCF of $(y + z)$. So

$$\begin{aligned}(ay+az)+(by+bz)&=a(y+z)+b(y+z) \quad \text{GCF of } (y+z)\\&=(y+z)(a+b)\end{aligned}$$

Example 2: Factor $x^2 + 2xy + y^2 - z^2$.

This polynomial has four terms with no common factor. Also notice the following:

$$x^2+2xy+y^2=(x+y)^2 \quad \text{square trinomial}$$

$$\begin{aligned} x^2+2xy+y^2-z^2 &= (x^2+2xy+y^2)-z^2 \\ &= (x+y)^2-z^2 \quad \text{difference of squares} \\ &= [(x+y)+z][(x+y)-z] \\ &= (x+y+z)\ (x+y-z) \end{aligned}$$

Example 3: Factor $x^2 - y^2 + x^3 + y^3$.

$$x^2-y^2+x^3+y^3=\underbrace{(x^2-y^2)}_{\substack{\text{difference}\\ \text{of squares}}}+\underbrace{(x^3+y^3)}_{\substack{\text{sum of}\\ \text{cubes}}}$$

$$=[(x+y)(x-y)]+[(x+y)(x^2-xy+y^2)]$$

The GFC is $(x+y)$.

$$\begin{aligned} &= (x+y)[(x-y)+(x^2-xy+y^2)] \\ &= (x+y)(x^2+x-xy-y+y^2) \end{aligned}$$

Summary of Factoring Techniques

- For all polynomials, first factor out the greatest common factor (GCF).
- For a binomial, check to see if it is any of the following:

  (a) difference of squares: $x^2-y^2=(x+y)(x-y)$

  (b) difference of cubes: $x^3-y^3=(x-y)(x^2+xy+y^2)$

  (c) sum of cubes: $x^3+y^3=(x+y)(x^2-xy+y^2)$

- For a trinomial, check to see whether it is either of the following forms:

  (a) $x^2 + bx + c$:

  If so, find two integers whose product is c and whose sum is b. For example,

  $$x^2 + 8x + 12 = (x + 2)(x + 6)$$

  since $(2)(6) = 12$ and $2 + 6 = 8$.

(b) $ax^2 + bx + c$:

If so, find two binomials so that
the product of first terms $= ax^2$,
the product of last terms $= c$, and
the sum of outer and inner products $= bx$.

See the following polynomial in which the product of the first terms $= (3x)(2x) = 6x^2$, the product of last terms $= (2)(-5) = -10$, and the sum of outer and inner products $= (3x)(-5) + 2(2x) = -11x$.

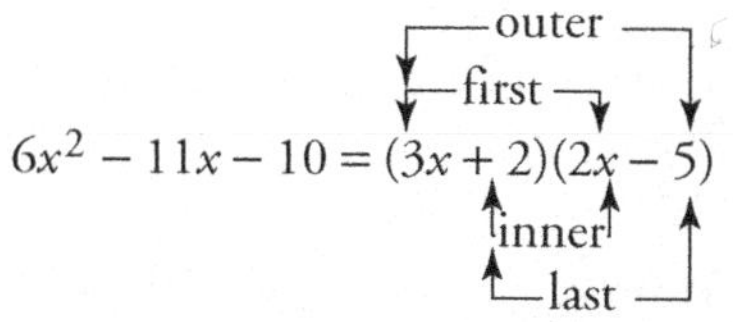

(c) $x^2 + 2xy + y^2 = (x + y)^2$ square trinomials

$x^2 - 2xy + y^2 = (x - y)^2$

- For polynomials with four or more terms, make groups, factor each group, and then find a pattern as in Example 1 in "Factoring by Grouping" on p. 119.

Solving Equations by Factoring

Factoring is a method that can be used to solve equations of a degree higher than 1. When an equation has one variable, there will be as many solutions as the degree of the polynomial. Recall that the degree is the largest exponent on the variable.

This method uses the zero-product rule.

> **Zero-Product Rule**
>
> If $(a)(b) = 0$, then either $a = 0$, $b = 0$, or both.
>
> Set the original polynomial equal to zero, factor the left side, then set each factor to zero and solve.

Example 1: Solve for x: $x(x + 3) = 0$.

$$x(x + 3) = 0$$

Apply the zero product rule.

$$x = 0 \text{ or } x + 3 = 0$$
$$x = 0 \text{ or } x = -3$$

Check the solution.

$$x(x+3) = 0 \qquad\qquad x(x+3) = 0$$
$$0(0+3) \stackrel{?}{=} 0 \qquad\qquad -3(-3+3) \stackrel{?}{=} 0$$
$$0(3) \stackrel{?}{=} 0 \qquad\qquad -3(0) \stackrel{?}{=} 0$$
$$0 = 0 \checkmark \qquad\qquad 0 = 0 \checkmark$$

The solution is $x = 0$ or $x = -3$. ***Note:*** There are two solutions because the original polynomial is of degree two.

Example 2: Solve for x: $x^2 - 5x + 6 = 0$.

$$x^2 - 5x + 6 = 0$$

Factor.

$$(x - 2)(x - 3) = 0$$

Apply the zero product rule.

$$x - 2 = 0 \quad \text{or} \quad x - 3 = 0$$
$$x = 2 \qquad\qquad x = 3$$

The solution is $x = 2$ or $x = 3$.

Example 3: Solve for x: $3x(2x - 5) = -4(4x - 3)$.

$$3x(2x - 5) = -4(4x - 3)$$

Distribute.

$$6x^2 - 15x = -16x + 12$$

In order to apply the zero product rule, you need to get all terms on one side, leaving zero on the other. Always remember to set the original polynomial equal to zero.

$$6x^2 + x - 12 = 0$$

Factor.

$$(3x - 4)(2x + 3) = 0$$

Apply the zero product rule.

$$\begin{aligned} 3x-4&=0 \quad \text{or} \quad & 2x+3&=0 \\ 3x&=4 & 2x&=-3 \\ x&=\frac{4}{3} & x&=-\frac{3}{2} \end{aligned}$$

The solution is $x=\frac{4}{3}$ or $x=-\frac{3}{2}$.

Example 4: Solve for y: $2y^3 = 162y$.

$$2y^3 = 162y$$

Get all terms on one side of the equation.

$$2y^3 - 162y = 0$$

Factor (GCF).

$$2y(y^2 - 81) = 0$$

Continue to factor (difference of squares).

$$2y(y + 9)(y - 9) = 0$$

Apply the zero product rule.

$$\begin{aligned} 2y&=0 \text{ or } & y+9&=0 \text{ or } & y-9&=0 \\ y&=0 & y&=-9 & y&=9 \end{aligned}$$

Note: There are three solutions because the original polynomial is of degree three.

The solution is $y = 0$ or $y = -9$ or $y = 9$.

Chapter Check-Out

Questions

1. Factor completely:

   (a) $36x^2 - 25$
   (b) $20x^2 - 80$

2. Factor:

   (a) $x^2 + 8x + 15$
   (b) $x^2 + 2x - 8$
   (c) $x^2 - 5x + 6$

3. Solve for x: $x^2 + 2x - 24 = 0$.
4. Solve for x: $6x^3 - x^2 - 35x = 0$.
5. Describe the hierarchy of factoring.
6. What does the degree of the original polynomial tell you?

Answers

1. **(a)** $(6x + 5)(6x - 5)$; **(b)** $20\ (x + 2)(x - 2)$
2. **(a)** $(x + 3)(x + 5)$; **(b)** $(x + 4)(x - 2)$; **(c)** $(x - 2)(x - 3)$
3. $x = 4$ or $x = -6$
4. $x = 0$ or $x = \frac{5}{2}$ or $x = -\frac{7}{3}$
5. Always look for the GCF and patterns $x^2 + bx + c$ and $ax^2 + bx + c$. (See "Summary of Factoring Techniques" on pp. 120–121.)
6. The degree of the original polynomial will tell the maximum number of solutions.

Chapter 7
RATIONAL EXPRESSIONS

Chapter Check-In

- ❑ Simplifying rational expressions
- ❑ Multiplying rational expressions
- ❑ Dividing rational expressions
- ❑ Adding and subtracting rational expressions
- ❑ Working with complex fractions
- ❑ Solving rational equations
- ❑ Working with proportions and variations
- ❑ Graphing rational functions

Common Core Standard: Seeing the Structure in Expressions

Rational expressions: addition, subtraction, multiplication, and division. Solve rational equations and graph rational functions. Arithmetic with polynomials and rational expressions (A.APR.3). Rewrite rational expressions in different forms (A.APR.6.7).

The term "rational expression" is a fancy way of saying *algebraic fraction.* A fraction also may be perceived as being a division problem, wherein the numerator is being divided by the denominator. Although the basic rules of arithmetic of fractions apply to the rational expressions treated within this chapter, having polynomials in the numerators and/or denominators does, at times, require some additional knowledge. This is even more evident when fractions are in the numerator and in the denominator (a complex fraction). This chapter also examines the concepts of direct, inverse, and joint variation; deals with solving rational equations; and explores graphing rational functions.

Examples of Rational Expressions

The quotient of two polynomials is a **rational expression.** The denominator of a rational expression will affect the domain values; it can *never* have a value of zero. The following are examples of rational expressions:

$$\frac{5}{6}, \frac{3x+8}{2x-9}, \frac{x^2-7x+3}{12}, \frac{9x}{x^2+5x+6}, 6x+5$$

The last example, $6x + 5$, could also be expressed as $\frac{6x+5}{1}$. Therefore, it satisfies the definition of a rational expression.

Simplifying Rational Expressions

To simplify a rational expression, follow these steps:

1. Completely factor the numerator and the denominator.
2. Reduce common factors.

Example 1: Simplify $\frac{5x+15}{4x+12}$.

$$\frac{5x+15}{4x+12} = \frac{5\overset{1}{\cancel{(x+3)}}}{4\underset{1}{\cancel{(x+3)}}} = \frac{5}{4}$$

Example 2: Simplify $\frac{x^2-16}{x^3+64}$.

$$\frac{x^2-16}{x^3+64} = \frac{\overset{1}{\cancel{(x+4)}}(x-4)}{\underset{1}{\cancel{(x+4)}}(x^2-4x+16)} = \frac{x-4}{x^2-4x+16}$$

Whenever possible, try to write all polynomials in descending order with a positive leading coefficient. To have a positive leading coefficient, occasionally -1 has to be factored out of the polynomial.

Example 3: Simplify $\frac{4x^2-25}{5-2x}$.

$$\frac{4x^2-25}{5-2x} = \frac{(2x+5)\overset{1}{\cancel{(2x-5)}}}{-1\underset{1}{\cancel{(2x-5)}}} = \frac{2x+5}{-1} = -2x-5$$

Example 4: Simplify $\dfrac{x^2-7x+12}{15-2x-x^2}$.

$$\frac{x^2-7x+12}{15-2x-x^2}=\frac{(x-3)(x-4)}{-1(x^2+2x-15)}=\frac{\overset{1}{\cancel{(x-3)}}(x-4)}{-1\underset{1}{\cancel{(x-3)}}(x+5)}=\frac{x-4}{-1(x+5)}$$

This answer can be expressed in other ways:

$$\frac{x-4}{-1(x+5)}=\frac{x-4}{-x-5},\quad \frac{x-4}{-1(x+5)}\cdot\frac{-1}{-1}=\frac{-x+4}{x+5},\text{ or } -\frac{x-4}{x+5}$$

Multiplying Rational Expressions

To multiply rational expressions, follow these steps:

1. Completely factor all numerators and denominators.
2. Reduce all common factors.
3. Either multiply the denominators and numerators or leave the answer in factored form.

Example 1: Multiply $\dfrac{x^2-8x+12}{x^2-16}\cdot\dfrac{4x+16}{x^2-4x+4}$.

$$\frac{x^2-8x+12}{x^2-16}\cdot\frac{4x+16}{x^2-4x+4}=\frac{\overset{1}{\cancel{(x-2)}}(x-6)}{\underset{1}{\cancel{(x+4)}}(x-4)}\cdot\frac{4\overset{1}{\cancel{(x+4)}}}{\underset{x-2}{\cancel{(x-2)^2}}}$$

$$=\frac{4(x-6)}{(x-4)(x-2)}$$

This last answer could be either left in its factored form or multiplied out. If multiplied out, it becomes

$$\frac{4x-24}{x^2-6x+8}$$

Example 2: Multiply $(x^2-2x)\cdot\dfrac{x}{x^2-5x+6}$.

$$(x^2-2x)\cdot\frac{x}{x^2-5x+6}=\frac{x(\overset{1}{\cancel{x-2}})}{1}\cdot\frac{x}{(\underset{1}{\cancel{x-2}})(x-3)}$$

$$=\frac{x^2}{x-3}$$

Example 3: Multiply $\dfrac{9-x^2}{x^2+6x+9}\cdot\dfrac{3x+9}{3x-9}$.

$$\frac{9-x^2}{x^2+6x+9}\cdot\frac{3x+9}{3x-9}=\frac{-1(\overset{1}{\cancel{x+3}})(\overset{1}{\cancel{x-3}})}{\underset{1}{\cancel{(x+3)^2}}}\cdot\frac{3(\overset{1}{\cancel{x+3}})}{3(\underset{1}{\cancel{x-3}})}$$

$$=\frac{-3}{3}=\frac{-1}{1}=-1$$

Dividing Rational Expressions

Just as in arithmetic, division of fractions involves multiplying by the reciprocal of the divisor.

Example 1: Divide $\dfrac{x^2-9x-10}{x^2+x-6}\div\dfrac{x^2-1}{x^2-4}$.

$$\frac{x^2-9x-10}{x^2+x-6}\div\frac{x^2-1}{x^2-4}=\frac{x^2-9x-10}{x^2+x-6}\cdot\frac{x^2-4}{x^2-1}$$

↑

Multiply by the reciprocal of $\dfrac{x^2-1}{x^2-4}$

$$\frac{x^2-9x-10}{x^2+x-6}\cdot\frac{x^2-4}{x^2-1}=\frac{(x-10)(\overset{1}{\cancel{x+1}})}{(\underset{1}{\cancel{x-2}})(x+3)}\cdot\frac{(x+2)(\overset{1}{\cancel{x-2}})}{(\underset{1}{\cancel{x+1}})(x-1)}$$

$$=\frac{(x-10)(x+2)}{(x+3)(x-1)}$$

If this last answer is multiplied out, it becomes $\dfrac{x^2-8x-20}{x^2+2x-3}$.

Example 2: Simplify $\dfrac{2x^2-2x-4}{x^2+2x-8} \div \dfrac{4x^2-100}{x^2-x-20} \cdot \dfrac{3x^2+15x}{x+1}$.

Since only the middle expression is doing the dividing, it is the only one whose reciprocal is used. Also recall that the division is performed first because it is on the left of the multiplication.

$$
\begin{aligned}
&\frac{2x^2-2x-4}{x^2+2x-8} \div \frac{4x^2-100}{x^2-x-20} \cdot \frac{3x^2+15x}{x+1} \\
&= \frac{2x^2-2x-4}{x^2+2x-8} \cdot \frac{x^2-x-20}{4x^2-100} \cdot \frac{3x^2+15x}{x+1} \\
&= \frac{\overset{1}{\cancel{2}}\overset{1}{(\cancel{x-2})}\overset{1}{(\cancel{x+1})}}{\underset{1}{(\cancel{x+4})}\underset{1}{(\cancel{x-2})}} \cdot \frac{\overset{1}{(\cancel{x+4})}\overset{1}{(\cancel{x-5})}}{\underset{2}{\cancel{4}}\underset{1}{(\cancel{x+5})}\underset{1}{(\cancel{x-5})}} \cdot \frac{3x\overset{1}{(\cancel{x+5})}}{\underset{1}{\cancel{x+1}}} \\
&= \frac{3x}{2}
\end{aligned}
$$

Adding and Subtracting Rational Expressions

To add or subtract rational expressions with the same denominators, take the following steps:

1. Add or subtract the numerators as indicated.
2. Keep the given denominator.
3. Simplify the resulting rational expression if possible.

Example 1: Add $\dfrac{4}{5x} + \dfrac{1}{5x}$.

$$\frac{4}{5x} + \frac{1}{5x} = \frac{4+1}{5x} = \frac{\overset{1}{\cancel{5}}}{\underset{1}{\cancel{5}}x} = \frac{1}{x}$$

Example 2: Simplify $\dfrac{x^2+5x+1}{x+3}-\dfrac{4x-5}{x+3}+\dfrac{7x+9}{x+3}$.

$$\begin{aligned}\frac{x^2+5x+1}{x+3}-\frac{4x-5}{x+3}+\frac{7x+9}{x+3}&=\frac{x^2+5x+1-(4x-5)+7x+9}{x+3}\\&=\frac{x^2+5x+1-4x+5+7x+9}{x+3}\\&=\frac{x^2+8x+15}{x+3}\\&=\frac{\overset{1}{\cancel{(x+3)}}(x+5)}{\underset{1}{\cancel{x+3}}}\\&=x+5\end{aligned}$$

Recall that the subtraction is performed first because it is on the left of the addition.

Adding and Subtracting Rational Expressions with Different Denominators

1. Completely factor each denominator.
2. Find the least common denominator (LCD) for all the denominators by multiplying together the different prime factors having the largest exponent for each factor.
3. Rewrite each fraction to have the LCD as its denominator by multiplying each fraction by the value of 1 in an appropriate form.
4. Combine numerators as indicated and keep the LCD as the denominator.
5. Simplify the resulting rational expression if possible.

Example 1: Add $\dfrac{5}{x}+\dfrac{2}{y}$.

Completely factor each denominator. x and y are already prime factors.

Find the least common denominator (LCD) for all the denominators. The LCD $= xy$.

Rewrite each fraction to have the LCD as its denominator.

$$\frac{5}{x}+\frac{2}{y}=\frac{5}{x}\cdot\left(\frac{y}{y}\right)+\frac{2}{y}\cdot\left(\frac{x}{x}\right)$$
$$=\frac{5y}{xy}+\frac{2x}{xy}$$

Combine numerators and keep the LCD as the denominator.

$$=\frac{5y+2x}{xy}$$

This rational expression cannot be simplified further.

Example 2: Add $\frac{4}{x^2-16}+\frac{3}{x^2+8x+16}$.

Factor each denominator.

$$x^2-16=(x+4)(x-4)$$
$$x^2+8x+16=(x+4)^2$$
$$\frac{4}{x^2-16}+\frac{3}{x^2+8x+16}=\frac{4}{(x+4)(x-4)}+\frac{3}{(x+4)^2}$$

Find the LCD. The LCD $= (x-4)(x+4)^2$.

Rewrite each fraction so that the LCD is its denominator.

$$\frac{4}{x^2-16}+\frac{3}{x^2+8x+16}=\frac{4}{(x+4)(x-4)}+\frac{3}{(x+4)^2}$$
$$=\frac{4}{(x+4)(x-4)}\cdot\left[\frac{(x+4)}{(x+4)}\right]+\frac{3}{(x+4)^2}\cdot\left[\frac{(x-4)}{(x-4)}\right]$$
$$=\frac{4x+16}{(x-4)(x+4)^2}+\frac{3x-12}{(x-4)(x+4)^2}$$

Combine numerators and keep the LCD as the denominator.

$$=\frac{7x+4}{(x-4)(x+4)^2}$$

This rational expression cannot be simplified further.

Example 3: Add $\dfrac{8x}{x-3}+\dfrac{5}{9-x^2}$.

Factor each denominator. Since $(x - 3)$ is a prime factor, move on to the second denominator and rewrite it in descending order.

$$9 - x^2 = -x^2 + 9$$

Factor out -1 so the leading coefficient is positive.

$$\begin{aligned}&=-1(x^2-9)\\&=-1(x-3)(x+3)\end{aligned}$$

The LCD $= (x - 3)(x + 3)$. [The LCD could also have been $-1(x - 3)(x + 3)$.]

$$\begin{aligned}\frac{8x}{(x-3)}+\frac{5}{-1(x-3)(x+3)}&=\frac{8x}{(x-3)}\cdot\left[\frac{(x+3)}{(x+3)}\right]+\frac{5}{-1(x-3)(x+3)}\cdot\left(\frac{-1}{-1}\right)\\&=\frac{8x^2+24x}{(x-3)(x+3)}+\frac{-5}{(x-3)(x+3)}\\&=\frac{8x^2+24x-5}{(x-3)(x+3)}\end{aligned}$$

This rational expression cannot be simplified further.

Example 4: Simplify $\dfrac{2x}{x^2-4}-\dfrac{1}{x^2-3x+2}+\dfrac{x+1}{x^2+x-2}$.

Factor each denominator.

$$\begin{aligned}x^2-4&=(x+2)(x-2)\\x^2-3x+2&=(x-2)(x-1)\\x^2+x-2&=(x+2)(x-1)\end{aligned}$$

The LCD $= (x + 2)(x - 2)(x - 1)$.

Rewrite the fraction so the LCD is the denominator.

$$\frac{2x}{x^2-4}-\frac{1}{x^2-3x+2}+\frac{x+1}{x^2+x-2}$$
$$=\frac{2x}{(x+2)(x-2)}-\frac{1}{(x-2)(x-1)}+\frac{x+1}{(x+2)(x-1)}$$
$$=\frac{2x}{(x+2)(x-2)}\cdot\left[\frac{(x-1)}{(x-1)}\right]-\frac{1}{(x-2)(x-1)}\cdot\left[\frac{(x+2)}{(x+2)}\right]+\frac{(x+1)}{(x+2)(x-1)}\cdot\left[\frac{(x-2)}{(x-2)}\right]$$
$$=\frac{2x^2-2x}{(x+2)(x-2)(x-1)}-\frac{x+2}{(x+2)(x-2)(x-1)}+\frac{x^2-2x+x-2}{(x+2)(x-2)(x-1)}$$
$$=\frac{2x^2-2x-x-2+x^2-2x+x-2}{(x+2)(x-2)(x-1)}$$
$$=\frac{3x^2-4x-4}{(x+2)(x-2)(x-1)}$$

Note: Subtraction was performed before addition because it was on the left of the addition.

This rational expression can be simplified.

$$=\frac{(3x+2)\overset{1}{\cancel{(x-2)}}}{(x+2)\underset{1}{\cancel{(x-2)}}(x-1)}$$
$$=\frac{(3x+2)}{(x+2)(x-1)}$$

Complex Fractions

If the numerator, denominator, or both contain fractions, then the expression is called a **complex fraction.** The fraction bar indicates division between the original numerator and the original denominator. To simplify a complex fraction, take the following steps:

1. Simplify both the numerator and the denominator expressions into single fraction expressions.
2. Divide the simplified numerator by the simplified denominator and simplify further if possible.

Example 1: Simplify $\dfrac{x+1-\dfrac{6}{x}}{\dfrac{1}{x}}$.

Simplify the numerator.

$$x+1-\frac{6}{x}=\frac{x^2+x-6}{x}$$

The denominator $\frac{1}{x}$ is already a single fraction expression, so it cannot be simplified. The complex fraction now becomes

$$\frac{\dfrac{x^2+x-6}{x}}{\dfrac{1}{x}}$$

Divide the simplified numerator by the denominator and simplify further.

$$=\frac{x^2+x-6}{x}\div\frac{1}{x}$$

$$=\frac{x^2+x-6}{x}\cdot\frac{x}{1}$$

$$=\frac{(x+3)(x-2)}{\cancel{x}_1}\cdot\frac{\cancel{x}^1}{1}$$

$$=(x+3)(x-2) \text{ or } x^2+x-6$$

Example 2: Simplify $\dfrac{\dfrac{1}{x^2}-\dfrac{1}{y^2}}{\dfrac{1}{x^3}+\dfrac{1}{y^3}}$.

Simplify both the numerator and the denominator into single fractions.

$$\frac{1}{x^2}-\frac{1}{y^2}=\frac{y^2-x^2}{x^2y^2}$$

$$\frac{1}{x^3}+\frac{1}{y^3}=\frac{y^3+x^3}{x^3y^3}$$

The complex fraction now becomes

$$\frac{\frac{y^2 - x^2}{x^2 y^2}}{\frac{y^3 + x^3}{x^3 y^3}}$$

Divide the simplified numerator by the simplified denominator and simplify further.

$$= \frac{y^2 - x^2}{x^2 y^2} \div \frac{y^3 + x^3}{x^3 y^3}$$

$$= \frac{\overbrace{y^2 - x^2}^{\text{difference of squares}}}{x^2 y^2} \cdot \frac{x^3 y^3}{\underbrace{y^3 + x^3}_{\text{sum of cubes}}}$$

$$= \frac{\overset{1}{\cancel{(y+x)}}(y-x)}{\underset{1}{\cancel{x^2 y^2}}} \cdot \frac{\overset{xy}{\cancel{x^3 y^3}}}{\underset{1}{\cancel{(y+x)}}(y^2 - xy + x^2)}$$

$$= \frac{xy(y-x)}{y^2 - xy + x^2}$$

Solving Rational Equations

An equation involving rational expressions is called a **rational equation.** To solve such an equation, follow these steps:

1. Completely factor all denominators.
2. Multiply both sides of the equation by the least common denominator (LCD). This step eliminates denominators from the entire equation.
3. Solve the resulting equation.
4. Eliminate any solutions that would make the LCD become zero. These solutions are called **extraneous solutions,** defined as solutions that do not make the original equation true.
5. Check the remaining solutions.

Example 1: Solve for x: $\frac{5}{x}+\frac{4}{x+3}=\frac{8}{x^2+3x}$.

Factor all denominators.

$$\frac{5}{x}+\frac{4}{x+3}=\frac{8}{x(x+3)} \qquad [\text{LCD}=x(x+3)]$$

Multiply both sides by the LCD.

$$x(x+3)\left[\frac{5}{x}+\frac{4}{x+3}\right]=x(x+3)\left[\frac{8}{x(x+3)}\right]$$

$$\frac{\overset{1}{\cancel{x}}(x+3)}{1}\cdot\frac{5}{\underset{1}{\cancel{x}}}+\frac{x(\overset{1}{\cancel{x+3}})}{1}\cdot\frac{4}{\underset{1}{\cancel{x+3}}}=\frac{\overset{1}{\cancel{x}}(\overset{1}{\cancel{x+3}})}{1}\cdot\frac{8}{\underset{1}{\cancel{x}}(\underset{1}{\cancel{x+3}})}$$

Solve the equation.

$$5x+15+4x=8$$
$$9x=-7$$
$$x=-\frac{7}{9}$$

Notice that $-\frac{7}{9}$ is not an extraneous solution because it does not make any original denominator zero.

Example 2: Solve for x: $\frac{x+1}{5}-2=\frac{-4}{x}$.

$$\frac{x+1}{5}-2=\frac{-4}{x} \qquad (\text{LCD}=5x)$$

Multiply both sides by the LCD.

$$5x\cdot\left(\frac{x+1}{5}-2\right)=5x\cdot\left(\frac{-4}{x}\right)$$

$$\frac{\overset{1}{\cancel{5}}x}{1}\cdot\left(\frac{x+1}{\underset{1}{\cancel{5}}}\right)-\frac{5x}{1}\cdot(2)=\frac{5\overset{1}{\cancel{x}}}{1}\cdot\left(\frac{-4}{\underset{1}{\cancel{x}}}\right)$$

Solve the equation.

$$x^2 + x - 10x = -20$$

Put all terms on one side and solve by factoring.

$$x^2 - 9x + 20 = 0$$
$$(x-4)(x-5) = 0$$
$$x = 4 \text{ or } x = 5$$

Neither answer is an extraneous solution.

Proportion, Direct Variation, Inverse Variation, Joint Variation

This section defines what proportion, direct variation, inverse variation, and joint variation are and explains how to solve such equations.

Proportion

A **proportion** is an equation stating that two rational expressions are equal. Simple proportions can be solved by applying the **cross products rule.**

> **Cross Products Rule**
> If $\frac{a}{b} = \frac{c}{d}$, then $ad = bc$.

More involved proportions are solved as rational equations.

Example 1: Solve for x: $\frac{x}{15} = \frac{7}{8}$.

Apply the cross products rule.

$$8x = (15)(7)$$
$$8x = 105$$
$$x = \frac{105}{8}$$

Example 2: Solve for x: $\frac{x+5}{3}=\frac{7x-2}{5}$.

Apply the cross products rule.

$$5(x+5)=3(7x-2)$$
$$5x+25=21x-6$$
$$-16x=-31$$
$$x=\frac{31}{16}$$

Example 3: Solve for x: $\frac{41x-12}{x^2-16}=\frac{4x+3}{x-4}$.

Multiply both sides by the LCD because it is easier than the cross products rule.

$$\frac{41x-12}{x^2-16}=\frac{4x+3}{x-4} \quad [\text{LCD}=(x+4)(x-4)]$$

$$\overset{1}{\cancel{(x+4)}}\,\overset{1}{\cancel{(x-4)}}\left[\frac{41x-12}{\underset{1}{\cancel{(x+4)}}\,\underset{1}{\cancel{(x-4)}}}\right]=(x+4)\,\overset{1}{\cancel{(x-4)}}\left[\frac{4x+3}{\underset{1}{\cancel{x-4}}}\right]$$
$$41x-12=4x^2+19x+12$$
$$0=4x^2-22x+24$$
$$0=2(2x^2-11x+12)$$
$$0=2(2x-3)(x-4)$$
$$2x-3=0 \text{ or } x-4=0$$
$$x=\frac{3}{2} \text{ or } x=4$$

However, $x = 4$ is an extraneous solution, because it makes the denominators of the original equation become zero. Therefore, $x=\frac{3}{2}$.

Direct variation

The phrase "y **varies directly** as x" or "y is directly proportional to x" means that as x gets bigger, so does y, and as x gets smaller, so does y. This concept of **direct variation** can be translated in two ways.

- $\frac{y}{x} = k$ for some constant k.

  The k is called the **constant of proportionality.** This translation is used when the constant is the desired result.

- $\frac{y_1}{x_1} = \frac{y_2}{x_2}$

  This translation is used when the desired result is either an original or a new value of x or y.

Example 4: If y varies directly as x, and $y = 10$ when $x = 7$, find the constant of proportionality.

Use $\frac{y}{x} = k$.

$$\frac{10}{7} = k$$

The constant of proportionality is $\frac{10}{7}$.

Example 5: If y varies directly as x, and $y = 10$ when $x = 7$, find y when $x = 12$.

Use $\frac{y_1}{x_1} = \frac{y_2}{x_2}$.

$$\frac{10}{7} = \frac{y}{12}$$

Apply the cross products rule.

$$7y = 120$$

$$y = \frac{120}{7}$$

Inverse variation

The phrase "y **varies inversely** as x" or "y is inversely proportional to x" means that as x gets bigger, y gets smaller, or vice versa. This concept of **inverse variation** can be translated in two ways.

- $yx = k$ for some constant k, called the constant of proportionality. Use this translation if the constant is desired.
- $y_1x_1 = y_2x_2$. Use this translation if a value of x or y is desired.

Example 6: If y varies inversely as x, and $y = 4$ when $x = 3$, find the constant of proportionality.

Use $yx = k$.

$$(4)(3) = k$$
$$k = 12$$

The constant of proportionality is 12.

Example 7: If y varies inversely as x, and $y = 9$ when $x = 2$, find y when $x = 3$.

Use $y_1x_1 = y_2x_2$.

$$(9)(2) = (y)(3)$$
$$18 = 3y$$
$$y = 6$$

Joint variation

If one variable varies as the product of other variables, it is called **joint variation.** The phrase "y varies jointly as x and z" can be translated in two ways.

- $\frac{y}{xz} = k$ if the constant is desired.
- $\frac{y_1}{x_1z_1} = \frac{y_2}{x_2z_2}$ if one of the variables is desired.

Example 8: If y varies jointly as x and z, and $y = 10$ when $x = 4$ and $z = 5$, find the constant of proportionality.

Use $\frac{y}{xz} = k$.

$$\frac{10}{(4)(5)} = k$$
$$\frac{10}{20} = k$$
$$k = \frac{1}{2}$$

Example 9: If y varies jointly as x and z, and $y = 12$ when $x = 2$ and $z = 3$, find y when $x = 7$ and $z = 4$.

Use $\frac{y_1}{x_1 z_1} = \frac{y_2}{x_2 z_2}$.

$$\frac{12}{(2)(3)} = \frac{y}{(7)(4)}$$
$$\frac{12}{6} = \frac{y}{28}$$
$$6y = 336$$
$$y = 56$$

Occasionally, a problem involves both direct and inverse variations. Suppose that y varies directly as x and inversely as z. This involves three variables and can be translated in two ways.

- $\frac{yz}{x} = k$ if the constant is desired.
- $\frac{y_1 z_1}{x_1} = \frac{y_2 z_2}{x_2}$ if one of the variables is desired.

Example 10: If y varies directly as x and inversely as z, and $y = 24$ when $x = 3$ and $z = 4$, find the constant of proportionality.

Use $\frac{yz}{x} = k$.

$$\frac{(24)(4)}{3} = k$$
$$\frac{96}{3} = k$$
$$32 = k$$

Example 11: If y varies directly as x and inversely as z, and $y = 5$ when $x = 2$ and $z = 4$, find y when $x = 3$ and $z = 6$.

Use $\frac{y_1 z_1}{x_1} = \frac{y_2 z_2}{x_2}$.

$$\frac{(5)(4)}{2} = \frac{(y)(6)}{3}$$
$$12y = 60$$
$$y = 5$$

Graphing Rational Functions

If $f(x)$ represents a rational expression, then $y = f(x)$ is a **rational function.** To graph a rational function, first find values for which the function is undefined. A function is undefined for any values that would make any denominator become zero. Dashed lines are drawn on the graph for any values for which the rational function is undefined. These lines are called vertical **asymptote lines.** The graph of the rational function will get close to these vertical asymptote lines, but it will never intersect them. Technically speaking, asymptotes show approachable but unattainable points and are not actually part of the graph of the function. They may be horizontal lines, vertical lines, or slant lines.

Example 1: Graph $y = \frac{2}{x+1}$.

First, notice that the function is undefined when $x = -1$ since that value of x makes the denominator equal to zero. So the graph of $x = -1$ becomes a vertical asymptote.

Second, find whether any horizontal asymptotes exist. To do this, solve the equation for x and see whether any values for y would make the new equation undefined. It may not always be easy to solve a rational function for x. For those cases, other methods are used involving the idea of limits, which is discussed in calculus.

To solve for x, apply the cross products rule.

$$\begin{aligned} y(x+1) &= 2 \\ xy + y &= 2 \\ xy &= 2 - y \\ x &= \frac{2-y}{y} \end{aligned}$$

This equation is undefined when $y = 0$. The graph of $y = 0$ becomes a horizontal asymptote. The graph of $x = -1$ and $y = 0$ is shown in the following figure.

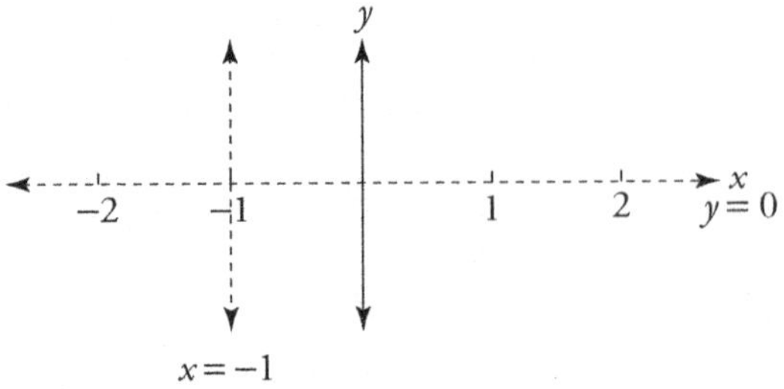

Third, plot points on each side of each asymptote line and use them to sketch the graph.

x	$y=\frac{2}{x+1}$	y
–4	$y=\frac{2}{-4+1}$	$-\frac{2}{3}$
–3	$y=\frac{2}{-3+1}$	–1
–2	$y=\frac{2}{-2+1}$	–2
–1	$y=\frac{2}{-1+1}$	undefined
0	$y=\frac{2}{0+1}$	2
1	$y=\frac{2}{1+1}$	1
2	$y=\frac{2}{2+1}$	$\frac{2}{3}$
3	$y=\frac{2}{3+1}$	$\frac{1}{2}$

Notice that graphs don't intersect asymptotes.

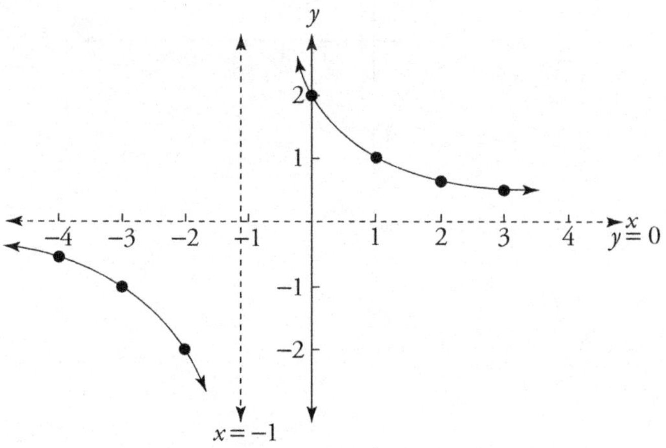

Example 2: Graph $y = \frac{x}{x+2}$.

First, notice that $x = -2$ is the vertical asymptote.

Second, find any horizontal asymptotes that exist by solving the equation for x. Apply the cross products rule.

$$\begin{aligned} y(x+2) &= x \\ xy + 2y &= x \\ xy - x &= -2y \\ x(y-1) &= -2y \\ x &= \frac{-2y}{y-1} \end{aligned}$$

This equation is undefined when $y = 1$, so $y = 1$ is a horizontal asymptote. The vertical and horizontal asymptotes are shown in the following graph.

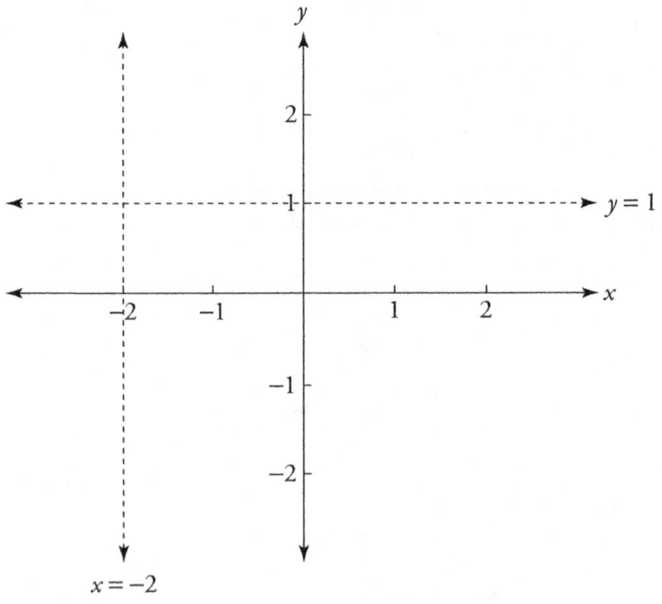

Third, plot points on either side of each asymptote line and sketch the graph.

x	$y=\frac{x}{x+2}$	y
–5	$y=\frac{-5}{-5+2}$	$\frac{5}{3}$
–4	$y=\frac{-4}{-4+2}$	2
–3	$y=\frac{-3}{-3+2}$	3
–2	$y=\frac{-2}{-2+2}$	undefined
–1	$y=\frac{-1}{-1+2}$	–1
0	$y=\frac{0}{0+2}$	0
1	$y=\frac{1}{1+2}$	$\frac{1}{3}$
2	$y=\frac{2}{2+2}$	$\frac{1}{2}$
3	$y=\frac{3}{3+2}$	$\frac{3}{5}$

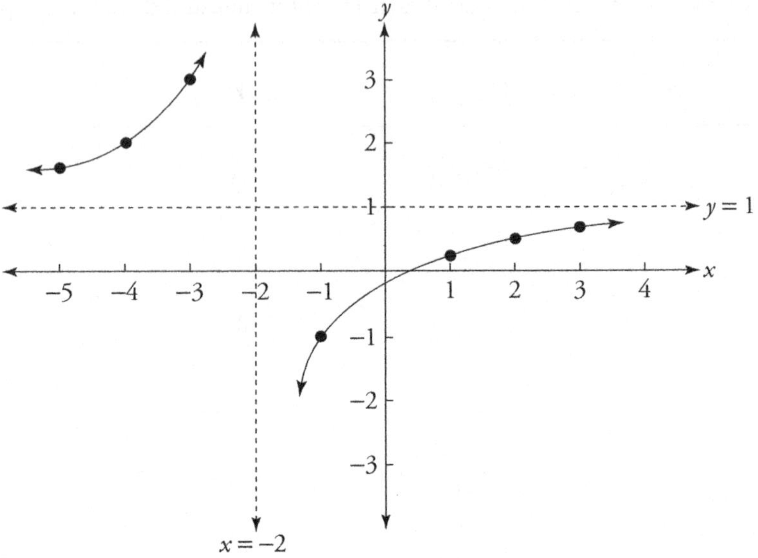

Chapter Check-Out

Questions

1. Simplify $\frac{9y^2 - 36}{6 - 3y}$.
2. Simplify $\frac{9}{x^2 - 9} + \frac{5}{x^2 + 6x + 9}$.
3. Solve for x: $\frac{x+7}{3} = \frac{3x-2}{5}$.
4. Solve for x: $\frac{x}{x^2 - 1} + \frac{2}{x+1} = \frac{1}{2x-2}$.
5. If y varies inversely as x, and $y = 6$ when $x = 3$, find the constant of proportionality.
6. Explain the difference between direct, inverse, and joint variations.
7. Explain how to find horizontal and vertical asymptotes when graphing rational functions.
8. A train travels 200 miles in the same amount of time that a car travels 300 miles. The rate of speed of the car is 20 mph more than the rate of speed of the train. Find the speed of the car and the train.

Answers

1. $-3y - 6$ or $-3(y + 2)$
2. $\dfrac{2(7x+6)}{(x+3)^2(x-3)}$
3. $x = \dfrac{41}{4}$
4. No solution.
5. $k = 18$
6. Answers may vary. In direct variation, the two values divide to a constant. In inverse variation, the two values multiply to a constant. Joint variation is a relationship between three values.
7. Answers may vary. Vertical asymptotes occur where the denominator is equal to zero. Horizontal asymptotes occur where the denominator is equal to zero, *after* the equation has been rewritten to solve for x.
8. The rate of speed of the car is 60 mph. The rate of speed of the train is 40 mph.

Chapter 8

RELATIONS AND FUNCTIONS

Chapter Check-In

- ❑ Defining relation, domain, range, and function
- ❑ Exploring function notation and composition of functions
- ❑ Learning the algebra of functions
- ❑ Creating inverse relations and inverse functions

Common Core Standard: Functions

Interpret functions that arise in applications in terms of context (F.IF). Build functions (F.BF). Construct and compare linear, quadratic, and exponential models and solve problems (F.LE). Create equations that describe numbers or relationships (A.CED).

In algebra, there is frequently more than a single name for something. That is also true of ordered pairs. The horizontal (or x) coordinates of a set of ordered pairs constitute the domain of a relation, while the range is determined by the vertical (or y) coordinates. A relation in which none of the domain numbers appears more than once is called a *function.* If two functions have a common domain, they can be acted upon arithmetically. This chapter also deals with the fact that the inverse of a function may not also be a function.

Basic Definitions

Following are definitions with which you should be familiar as you work with relations and functions.

Relation

A **relation** is a set of pairs (values) that can be represented by ordered pairs, a diagram, a mapping, a graph, or a mathematical sentence (equation) that relates the value of one variable to that of the other.

Domain and range

The set of all first numbers of the ordered pairs in a relation is called the **domain** of the relation. The set of all second numbers of the ordered pairs in a relation is called the **range** of the relation. The values in the domain and range are usually listed from least to greatest.

Example 1: Find the domain and range for the following relation:

$\{(4, 1), (3, 2), (-1, 6)\}$.

domain $= \{-1, 3, 4\}$
range $= \{1, 2, 6\}$

Example 2: Find the domain and range for the following relation:

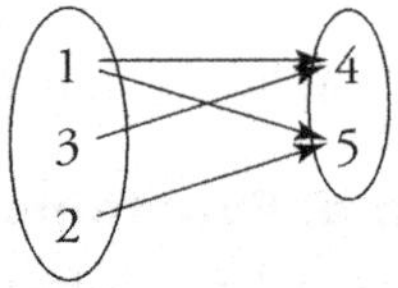

domain $= \{1, 2, 3\}$
range $= \{4, 5\}$

Example 3: Find the domain and range for the following relation:

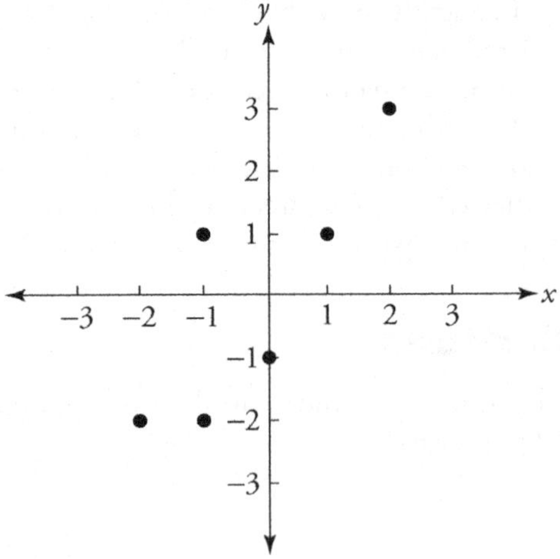

domain $= \{-2, -1, 0, 1, 2\}$
range $= \{-2, -1, 1, 3\}$

Example 4: Find the domain and range for the following relation:

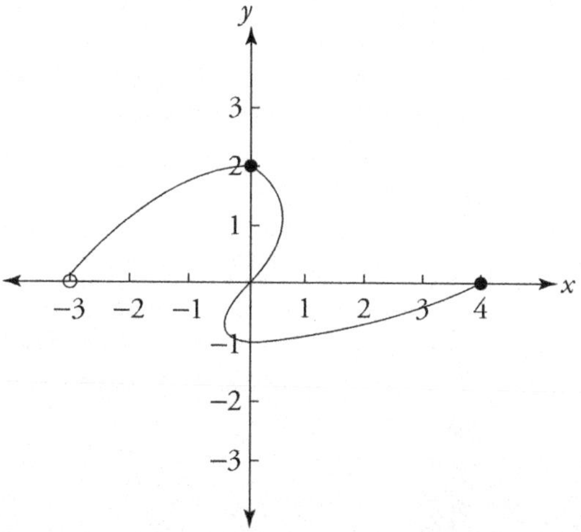

The domain and range cannot be listed as in the preceding examples. In order to visualize the domain, imagine each point of the graph going vertically to the *x*-axis. The points on the *x*-axis become the horizontal domain, as shown in the following figure.

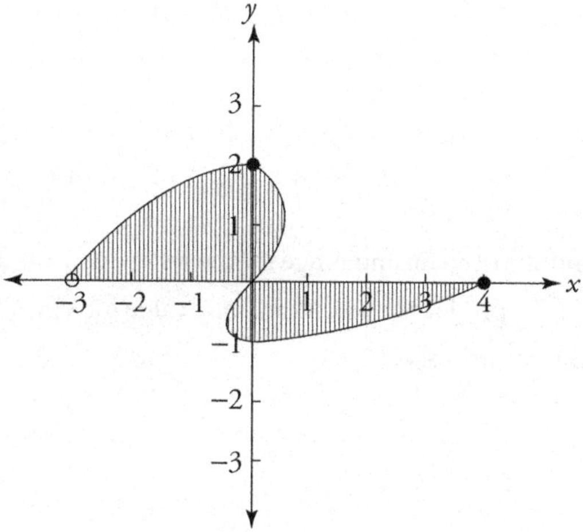

(a)

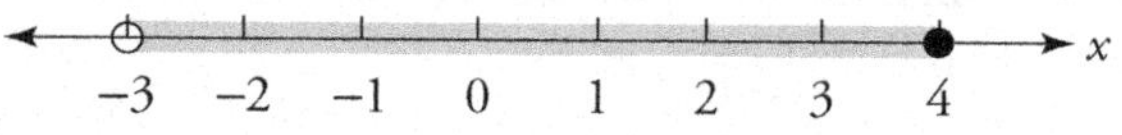

(b)

The domain can be expressed as domain $= \{x|-3 < x \leq 4\}$, which is read as "the set of all values of x, such that x is greater than -3 and x is less than or equal to 4."

To visualize the range, have all the points move horizontally to the y-axis. The points on the y-axis become the vertical range, as shown in the following figure.

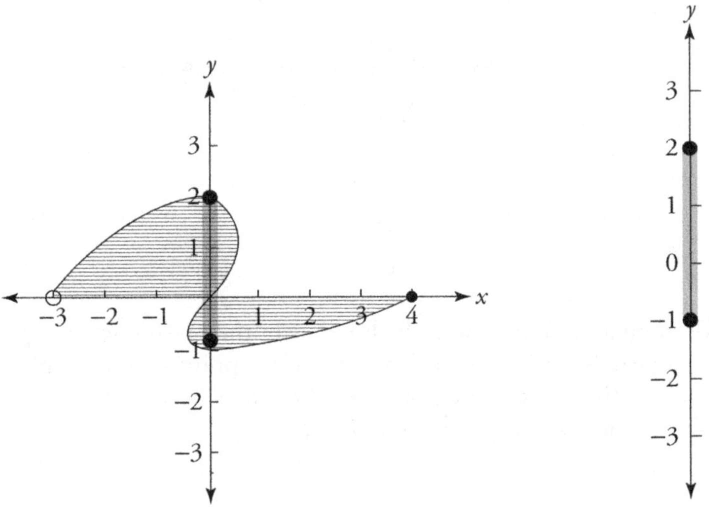

(a) (b)

The range can be expressed as range $= \{y|-1 \leq y \leq 2\}$, which is read as "the set of all values of y, such that y is greater than or equal to -1 and y is less than or equal to 2."

Example 5: Find the domain and range for the following relation: $y = 2x + 3$.

Since any value for x produces a y value and any value for y produces an x value,

domain = {all real numbers}
range = {all real numbers}

Function

A relation in which none of the domain values are repeated is called a **function.**

Example 6: Which of the following relations are functions?

(a) $\{(4, 1), (3, 2), (-1, 6)\}$

This relation is a function since domain values are not repeated.

(b)

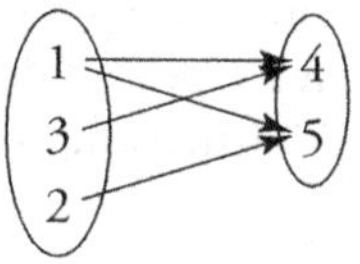

This relation is not a function. The diagram represents the set of ordered pairs {(1, 4), (1, 5), (3, 4), (2, 5)}. The pairs (1, 4) and (1, 5) repeat domain values, making it not a function.

(c)

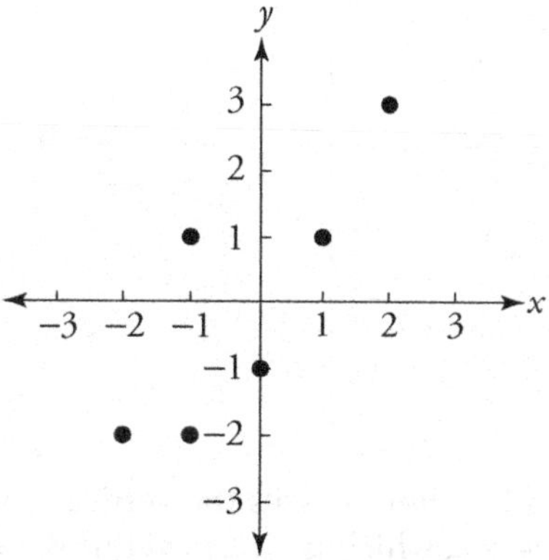

This relation is not a function. The graph represents the set of ordered pairs {(−2, −2), (−1, −2), (−1, 1), (0, −1), (1, 1), (2, 3)}. The ordered pairs (−1, −2) and (−1, 1) repeat domain values, making it not a function.

Notice that the vertical line $x = -1$ would pass through two points of the graph, as shown in the following figure.

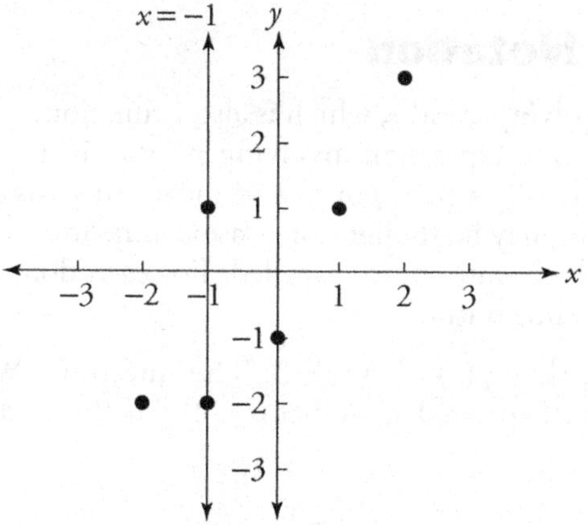

A vertical line that passes through (intersects) a graph in more than one point indicates ordered pairs that have repeated domain values and eliminates the relation from being called a function. This test for functions is known as the **vertical line test.**

(d)

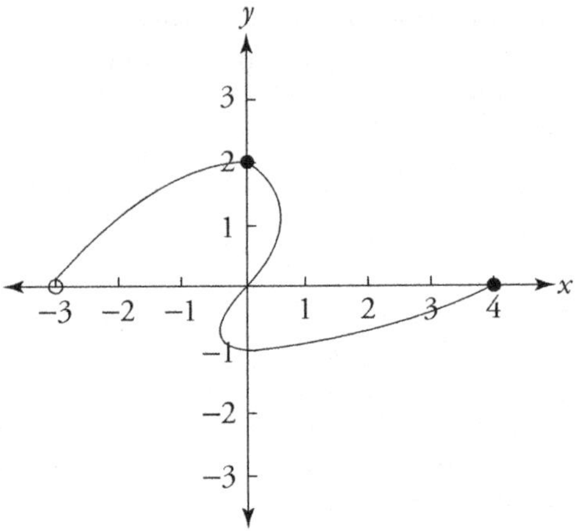

This relation is not a function. It fails the vertical line test. To illustrate this, notice that the y-axis, which is a vertical line drawn at $x = 0$, crosses the relation at three points: (0, 2), (0, 0), and (0, −1).

(e) $y = 2x + 3$

This relation is a function, since domain values are not repeated. When $y = 2x + 3$ is graphed, it can be seen that the graph passes the vertical line test.

Function Notation

An equation involving x and y, which is also a function, can be written in the form y = "some expression involving x"; that is, $y = f(x)$. This last expression is read as "y equals f of x" and means that y is a function of x. This concept also may be thought of as a machine into which inputs are fed and from which outputs are expelled. For each domain value input, there is a range value output.

If $y = 4x + 3$, then $f(x) = 4x + 3$. The question "What is y when $x = 2$?" would be expressed as "What is $f(2)$?" if $f(x) = 4x + 3$.

$$\begin{aligned} f(2) &= 4(2) + 3 \\ &= 11 \end{aligned}$$

The statement $f(2) = 11$ says, "When x equals 2, y equals 11"; that is, the ordered pair (2, 11) belongs to this function. That also may be read as "when a 2 is input into this function, an 11 is output."

Example 1:

(a) If $f(x) = x^2 + 2x + 3$, find $f(-3)$.

$$\begin{aligned} f(x) &= x^2 + 2x + 3 \\ f(-3) &= (-3)^2 + 2(-3) + 3 \\ &= 9 - 6 + 3 \\ &= 6 \end{aligned}$$

(b) If $f(x) = x^2 + 2x + 3$, find $f(0)$.

$$\begin{aligned} f(x) &= x^2 + 2x + 3 \\ f(0) &= (0)^2 + 2(0) + 3 \\ &= 3 \end{aligned}$$

Composition of Functions

When the input in a function is another function, the result is called a **composite function.** For example, if $f(x) = 3x + 2$ and $g(x) = 4x + 5$, then $f[g(x)]$ is a composite function. The statement $f[g(x)]$ is read "f of g of x" or "the composition of f with g." $f[g(x)]$ can also be written as

$$(f \circ g)(x) \text{ or } f \circ g(x)$$

The symbol between f and g is a small open circle. When replacing one function with another, be very careful to get the order correct because compositions of functions are not necessarily commutative, as you'll see in the example problems that follow.

Example 1: If $f(x) = 3x + 2$ and $g(x) = 4x - 5$, find each of the following.

(a) $f[g(4)]$

$$\begin{aligned} f(x) &= 3x + 2 \\ f[g(4)] &= 3[g(4)] + 2 \\ &= 3(11) + 2 \\ &= 35 \end{aligned}
\qquad
\left[\begin{aligned} g(x) &= 4x - 5 \\ g(4) &= 4(4) - 5 \\ &= 16 - 5 \\ &= 11 \end{aligned}\right]$$

(b) $g \circ f(4)$

$$\begin{aligned} g(x) &= 4x - 5 \\ g \circ f(4) &= 4[f(4)] - 5 \\ &= 4(14) - 5 \\ &= 51 \end{aligned} \qquad \left[\begin{aligned} f(x) &= 3x + 2 \\ f(4) &= 3(4) + 2 \\ &= 12 + 2 \\ &= 14 \end{aligned}\right]$$

(c) $f[g(x)]$

$$\begin{aligned} f(x) &= 3x + 2 \\ f[g(x)] &= 3[g(x)] + 2 \\ &= 3(4x - 5) + 2 \\ &= 12x - 15 + 2 \\ &= 12x - 13 \end{aligned} \qquad [g(x) = 4x - 5]$$

(d) $(g \circ f)(x)$

$$\begin{aligned} g(x) &= 4x - 5 \\ g \circ f(x) &= 4[f(x)] - 5 \\ &= 4(3x + 2) - 5 \\ &= 12x + 8 - 5 \\ &= 12x + 3 \end{aligned} \qquad [f(x) = 3x + 2]$$

From the previous example, you should be convinced that $f[g(x)]$ and $g[f(x)]$ are not (necessarily) equal.

Example 2: If $f(x) = 3x^2 + 2x + 1$ and $g(x) = 4x - 5$, find each of the following.

(a) $f[g(x)]$

$$\begin{aligned} f(x) &= 3x^2 + 2x + 1 \qquad [g(x) = 4x - 5] \\ f[g(x)] &= 3[g(x)]^2 + 2[g(x)] + 1 \\ &= 3(4x - 5)^2 + 2(4x - 5) + 1 \\ &= 3(16x^2 - 40x + 25) + 8x - 10 + 1 \\ &= 48x^2 - 120x + 75 + 8x - 10 + 1 \\ &= 48x^2 - 112x + 66 \end{aligned}$$

(b) $g[f(x)]$

$$g(x) = 4x - 5 \qquad [f(x) = 3x^2 + 2x + 1]$$
$$\begin{aligned} g[f(x)] &= 4[f(x)] - 5 \\ &= 4(3x^2 + 2x + 1) - 5 \\ &= 12x^2 + 8x + 4 - 5 \\ &= 12x^2 + 8x - 1 \end{aligned}$$

Algebra of Functions

If two functions have a common domain, then arithmetic can be performed with them using the following definitions.

$(f + g)(x) = f(x) + g(x)$

$(f - g)(x) = f(x) - g(x)$

$(f \cdot g)(x) = f(x)g(x)$

$\left(\dfrac{f}{g}\right)(x) = \dfrac{f(x)}{g(x)}$, where $g(x) \neq 0$

Example 1: If $f(x) = x + 4$ and $g(x) = x^2 - 2x - 3$, find each of the following and determine the common domain.

(a) $(f + g)(x)$

$$\begin{aligned} (f+g)(x) &= (x+4)+(x^2-2x-3) \\ &= x^2-x+1 \end{aligned}$$

The common domain is {all real numbers}.

(b) $(f - g)(x)$

$$\begin{aligned} (f-g)(x) &= (x+4)-(x^2-2x-3) \\ &= x+4-x^2+2x+3 \\ &= -x^2+3x+7 \end{aligned}$$

The common domain is {all real numbers}.

(c) $(f \cdot g)(x)$

$$(f \cdot g)(x) = (x + 4)(x^2 - 2x - 3)$$

Use the distributive property.

$$= x^3 - 2x^2 - 3x + 4x^2 - 8x - 12$$
$$= x^3 + 2x^2 - 11x - 12$$

The common domain is {all real numbers}.

(d) $\left(\frac{f}{g}\right)(x)$

$$\left(\frac{f}{g}\right)(x) = \frac{x+4}{x^2 - 2x - 3}$$
$$= \frac{x+4}{(x-3)(x+1)}$$

This expression is undefined when $x = 3$ or when $x = -1$. So the common domain is {all real numbers except 3 or -1}.

Inverse Relations and Inverse Functions

If the ordered pairs of a relation, R, are reversed, then the new set of ordered pairs is called the **inverse relation** of the original relation.

Example 1: If $R = \{(1, 2), (3, 8), (5, 6)\}$, find the inverse relation of R. (The inverse relation of R is written R^{-1}. ***Note:*** It does not have the same meaning as a negative exponent.

$$R^{-1} = \{(2, 1), (8, 3), (6, 5)\}$$

Notice that the domain of R^{-1} is the range of R, and the range of R^{-1} is the domain of R. If a relation and its inverse are graphed, they will be symmetrical about the line $y = x$.

Example 2: Graph R and R^{-1} from Example 1, along with the line $y = x$ on the same set of coordinate axes.

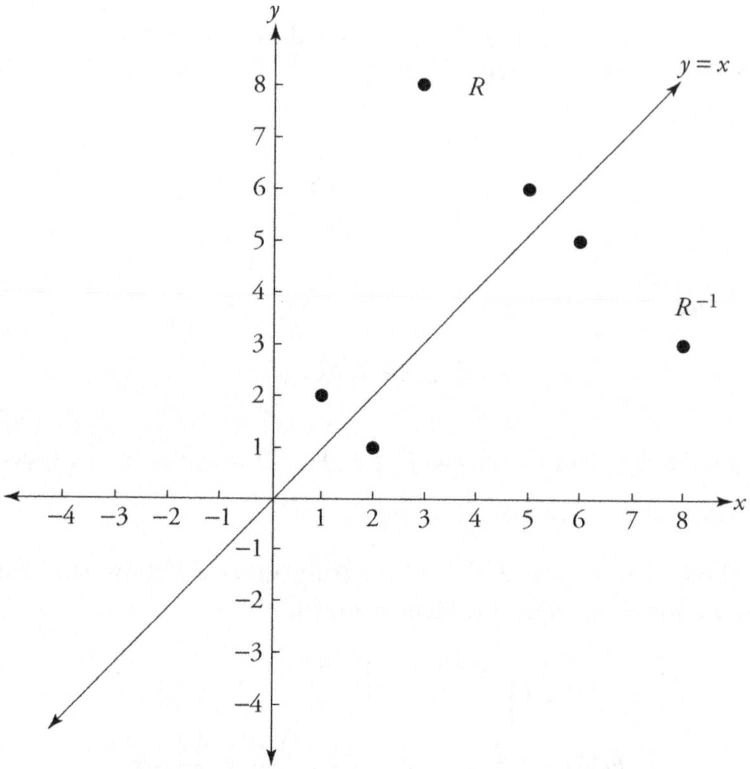

If this graph were "folded over" the line $y = x$, the set of points called R would coincide with the set of points called R^{-1}, making the two sets symmetrical about the line $y = x$.

As with inverse relations, there are also **inverse functions.** Two functions, f and g, are inverses of each other when the composition $f[g(x)]$ and $g[f(x)]$ are both the identity function. That is, $f[g(x)] = g[f(x)] = x$. The **identity function** is defined as the function $y = x$, or $f(x) = x$; for each replacement of x, the result is identical to x.

Example 3: If $f(x) = 4x - 5$, find $f^{-1}(x)$.

$f(x) = 4x - 5$ means $y = 4x - 5$. $f^{-1}(x)$ is the notation used for the inverse of the $f(x)$ function.

To find $f^{-1}(x)$, simply interchange the x and y variables and solve for the "new" y. The original equation is $y = 4x - 5$. The inverse is $x = 4y - 5$.

Solve for y.

$$4y - 5 = x$$

$$4y = x + 5$$

$$y = \frac{x+5}{4}$$

Therefore,

$$f^{-1} = \frac{x+5}{4}$$

For any ordered pair that makes $f(x) = 4x - 5$ true, the reverse ordered pair will make $f^{-1}(x) = \frac{(x+5)}{4}$ true.

To show that $f(x)$ and $f^{-1}(x)$ are truly inverses, show that their compositions both equal the identity function.

$$f[f^{-1}(x)] = f\left[\frac{x+5}{4}\right] = 4\left(\frac{x+5}{4}\right) - 5 = x + 5 - 5 = x$$

$$f^{-1}[f(x)] = f^{-1}[4x-5] = \frac{(4x-5)+5}{4} = \frac{4x}{4} = x$$

Since $f[f^{-1}(x)] = f^{-1}[f(x)] = x$, then $f(x)$ and $f^{-1}(x)$ are inverses of each other.

Example 4: Graph $f(x)$ and $f^{-1}(x)$ from Example 3 together with the identity function on the same set of coordinate axes.

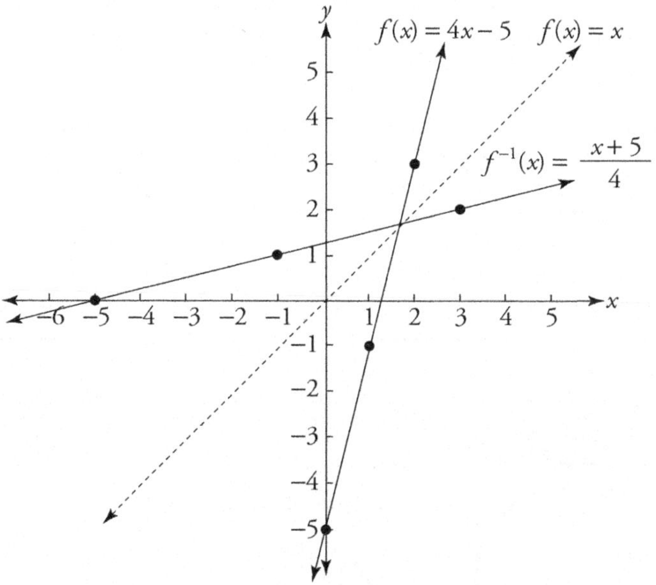

Notice that if the graph were "folded over" the identity function, the graphs of $f(x)$ and $f^{-1}(x)$ would coincide.

Example 5: If $f(x) = x^2$, find $f^{-1}(x)$.

$f(x) = x^2$ means $y = x^2$. The original equation is $y = x^2$. The inverse is $x = y^2$.

Solve for y.

$$y^2 = x$$
$$y = \pm\sqrt{x}$$

There are two relations for $f^{-1}(x)$: $f^{-1}(x) = \sqrt{x}$ and $f^{-1}(x) = -\sqrt{x}$.

In order for both $f(x)$ and $f^{-1}(x)$ to be functions, a restriction needs to be made on the domain of $f(x)$ so only one relation appears as $f^{-1}(x)$. If the domain of $f(x)$ is restricted to $\{x|x \geq 0\}$, then $f^{-1}(x) = \sqrt{x}$ is the only answer for $f^{-1}(x)$. If the domain of $f(x)$ is restricted to $\{x|x \leq 0\}$, then $f^{-1}(x) = -\sqrt{x}$ is the only answer for $f^{-1}(x)$.

Example 6: Graph $f(x) = x^2$ together with $f^{-1}(x) = \sqrt{x}$, $f^{-1}(x) = -\sqrt{x}$, and the identity function $f(x) = x$ all on the same set of coordinate axes.

To graph $f(x) = x^2$, find several ordered pairs that make the sentence $y = x^2$ true. To graph $f^{-1}(x) = \pm\sqrt{x}$, simply take the reverse of the ordered pairs found for $f(x) = x^2$.

x	$f(x) = x^2$
–3	9
–2	4
–1	1
0	0
1	1
2	4
3	9

x	$f^{-1}(x) = \sqrt{x}$
9	3
4	2
1	1
0	0

x	$f^{-1}(x) = -\sqrt{x}$
9	–3
4	–2
1	–1
0	0

This is an example of a function whose inverse is only a relation.

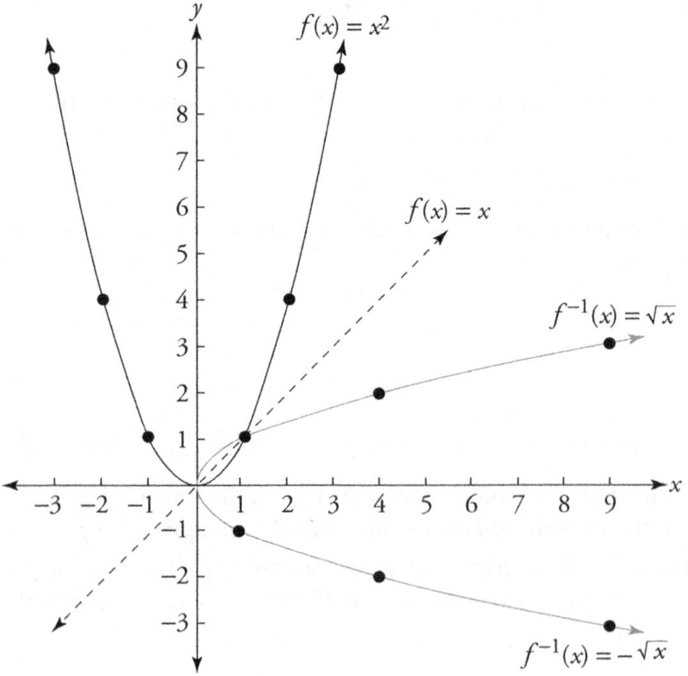

Notice that $f(x) = x^2$ is a function. From the tables, you can see that $f^{-1}(x) = \sqrt{x}$ is a function by itself, and so is $f^{-1}(x) = -\sqrt{x}$. But together, $f^{-1}(x) = \pm\sqrt{x}$ is not a function. The reason is that $f^{-1}(x) = \pm\sqrt{x}$ does not pass the vertical line test. Also notice that $f(x)$ and $f^{-1}(x)$ will coincide when the graph is "folded over" the identity function. Thus, the two relations are inverses of each other.

Example 7: Graph $f(x) = x^2$ with the restricted domain $\{x|x \geq 0\}$ together with $f^{-1}(x) = \sqrt{x}$ and the identity function on the same set of coordinate axes.

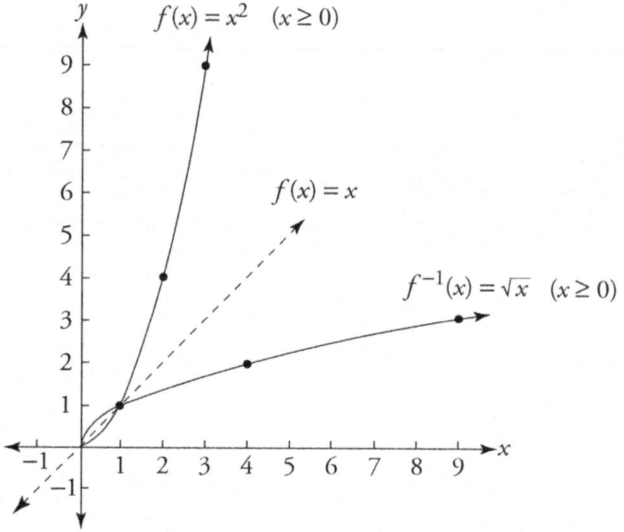

Notice that $f(x)$ and $f^{-1}(x)$ are now both functions, and they are symmetrical with respect to $f(x) = x$. To show that $f(x) = x^2$ and $f^{-1}(x) = \sqrt{x}$ are inverse functions, show that their compositions each produce the identity function.

$$f\left[f^{-1}(x)\right] = f\left[\sqrt{x}\right] = \left(\sqrt{x}\right)^2 = x$$

$$f^{-1}\left[f(x)\right] = f^{-1}\left[x^2\right] = \sqrt{x^2} = x \qquad \text{(since } x \geq 0\text{)}$$

Chapter Check-Out

Questions

1. If $f(x) = x^2 + 7x + 5$, find the following:
   (a) $f(7)$
   (b) $f(-3)$
   (c) $f(0)$
2. If $f(x) = 3x - 7$, find $f^{-1}(x)$. Show that they are inverse functions.
3. If $f(x) = x + 5$ and $g(x) = x^2 - 4x - 5$, find the following:
   (a) $(f + g)(x)$
   (b) $(f \cdot g)(x)$
   (c) $\frac{g(x)}{f(x)}$
4. Explain the difference between a relation and a function.
5. When is it necessary to restrict a domain?
6. Describe finding an inverse function graphically and algebraically.

Answers

1. (a) 103; (b) -7; (c) 5
2. $f^{-1}(x) = \frac{x+7}{3}$

$$3\left(\frac{x+7}{3}\right) - 7 = x$$
$$x + 7 - 7 = x$$
$$x = x$$

3. (a) $x^2 - 3x$; (b) $x^3 + x^2 - 25x - 25$; (c) $\frac{x^2 - 4x - 5}{x+5}$, where $x \neq -5$
4. All functions are relations, but not all relations are functions. A function is a special relation whose domain never has repeated values.
5. A function whose denominator contains an algebraic expression has to exclude the values of the variable that would produce a value of zero in the denominator.
6. Graphically, a function and its inverse will be symmetrical about the line $y = x$. Algebraically, interchange the x and y variables and solve for the "new" y.

Chapter 9
POLYNOMIAL FUNCTIONS

Chapter Check-In

- ❑ Defining a polynomial function
- ❑ Applying the remainder theorem
- ❑ Using the factor theorem
- ❑ Testing for zeros and rational zeros
- ❑ Graphing polynomial functions

Common Core Standard: Seeing the Structure in Expressions

Solve and graph polynomial functions. Arithmetic with polynomials and rational expressions. Understand the relationship between zeros and factors of polynomials. Use polynomial identities to solve problems. Rewrite rational expressions (A.APR.2-7).

This chapter deals with polynomial functions, showing how to evaluate them and how much can be determined about them by using the factor theorem, the remainder theorem, and their zeros. A zero may be any replacement for the variable that will result in the function having a value of zero, or any x value that produces a y value of zero; also known as an x-intercept on a graph. Graphing of polynomial functions is also dealt with in detail.

Polynomial Function

A **polynomial function** is any function of the form

$$P(x) = a_0x^n + a_1x^{n-1} + a_2x^{n-2} + \ldots + a_{n-1}x + a_n$$

where the coefficients $a_0, a_1, a_2, \ldots, a_n$ are real numbers and n is a whole number. Polynomial functions are evaluated by replacing the variable with a value. The instruction "evaluate the polynomial function $P(x)$ when x is replaced with 4" is written as "find $P(4)$."

Example 1: If $P(x) = 3x^3 - 2x^2 + 5x + 3$, find $P(-4)$.

$$\begin{aligned} P(-4) &= 3(-4)^3 - 2(-4)^2 + 5(-4) + 3 \\ &= 3(-64) - 2(16) - 20 + 3 \\ &= -192 - 32 - 20 + 3 \\ &= -241 \end{aligned}$$

Example 2: If $f(x) = 3x^2 - 4x + 5$, find $f(x + h)$.

$$\begin{aligned} f(x+h) &= 3(x+h)^2 - 4(x+h) + 5 \\ &= 3(x^2 + 2xh + h^2) - 4x - 4h + 5 \\ &= 3x^2 + 6xh + 3h^2 - 4x - 4h + 5 \end{aligned}$$

Remainder Theorem

If a polynomial $P(x)$ is divided by $(x - r)$, then the remainder of this division is the same as evaluating $P(r)$, and evaluating $P(r)$ for some polynomial $P(x)$ is the same as finding the remainder of $P(x)$ divided by $(x - r)$. This is known as the remainder theorem.

Example 1: Find $P(-3)$ if $P(x) = 7x^5 - 4x^3 + 2x - 11$.

There are two methods of finding $P(-3)$.

- **Method 1:** Directly replace -3 for x.
- **Method 2:** Find the remainder when $P(x)$ is divided by $[x - (-3)]$.

Method 1:

$$\begin{aligned} P(x) &= 7x^5 - 4x^3 + 2x - 11 \\ P(-3) &= 7(-3)^5 - 4(-3)^3 + 2(-3) - 11 \\ &= 7(-243) - 4(-27) - 6 - 11 \\ &= -1{,}701 + 108 - 6 - 11 \\ &= -1{,}610 \end{aligned}$$

Method 2:

$P(x)$ divided by $[x - (-3)]$ will be done by synthetic division (as explained in Chapter 5).

$$\begin{array}{r|rrrrrr} -3 & 7 & 0 & -4 & 0 & 2 & -11 \\ & & -21 & 63 & -177 & 531 & -1{,}599 \\ \hline & 7 & -21 & 59 & -177 & 533 & -1{,}610 \end{array}$$

Therefore, $P(-3) = -1{,}610$.

Example 2: Find the remainder when $P(x)$ is divided by $(x - 4)$ if $P(x) = x^4 + x^3 - 13x^2 - 25x - 12$.

Method 1:

$$\begin{aligned} \text{remainder} &= P(4) \\ &= (4)^4 + (4)^3 - 13(4)^2 - 25(4) - 12 \\ &= 256 + 64 - 208 - 100 - 12 \\ &= 0 \end{aligned}$$

Method 2:

$$\begin{array}{r|rrrrr} 4 & 1 & 1 & -13 & -25 & -12 \\ & & 4 & 20 & 28 & 12 \\ \hline & 1 & 5 & 7 & 3 & 0 \end{array}$$

Therefore, the remainder $= 0$.

In Example 2, since the division has a remainder of zero, both the **divisor** (the expression doing the dividing) and the **quotient** (the answer) are **factors** of the **dividend** (the expression being divided).

Factor Theorem

If $P(x)$ is a polynomial, then $P(r) = 0$ if and only if $x - r$ is a factor of $P(x)$.

Example 1: Is $(x + 2)$ a factor of $x^3 - x^2 - 10x - 8$?

Check to see whether $(x^3 - x^2 - 10x - 8) \div (x + 2)$ has a remainder of zero. Using synthetic division, you get

$$\begin{array}{r|rrrr} -2 & 1 & -1 & -10 & -8 \\ & & -2 & 6 & 8 \\ \hline & 1 & -3 & -4 & \boxed{0} \end{array} \leftarrow \text{remainder}$$

Because the remainder of the division is zero, $(x + 2)$ is a factor of $x^3 - x^2 - 10x - 8$. The expression $x^3 - x^2 - 10x - 8$ can now be expressed in factored form. This expression is referred to as the *depressed polynomial.*

$$x^3 - x^2 - 10x - 8 = (x + 2)(x^2 - 3x - 4)$$

Then we need to see if the depressed polynomial can still be factored.

$(x^2 - 3x - 4)$ can be factored further into $(x - 4)(x + 1)$. Therefore,

$$x^3 - x^2 - 10x - 8 = (x + 2)(x - 4)(x + 1)$$

The expression $x^3 - x^2 - 10x - 8$ is now **completely factored.** From this form, it also is seen that $(x - 4)$ and $(x + 1)$ are also factors of $x^3 - x^2 - 10x - 8$.

Zeros of a Function

The **zero of a function** is any replacement for the variable that will produce an answer of zero. Graphically, the real zero of a function is where the graph of the function crosses the x-axis; that is, the real zero of a function is the x-intercept(s) of the graph of the function. Algebraically, the zeros of the function are the x values that solve the equation when it is set equal to zero.

Example 1: Find the zeros of the function $f(x) = x^2 - 8x - 9$.

Find x such that $f(x) = 0$. The zeros of the function $f(x) = x^2 - 8x - 9$ are -1 and 9. This means $f(-1) = 0$ and $f(9) = 0$.

Note: These values can also be referred to as the ordered pairs $(-1, 0)$ and $(9, 0)$.

If a polynomial function with integer coefficients has real zeros, then they are either rational or irrational values. Rational zeros can be found by using the rational zero theorem.

Rational Zero Theorem

If a polynomial function, written in descending order of the exponents, has integer coefficients, then any rational zero must be of the form $\pm\frac{p}{q}$, where p is a factor of the constant term and q is a factor of the leading coefficient.

Example 1: Find all of the possible rational zeros of $f(x) = 2x^3 + 3x^2 - 8x + 3$.

According to the rational zero theorem, any rational zero must have a factor of 3 in the numerator and a factor of 2 in the denominator.

$$p: \text{factors of } 3 = \pm 1, \pm 3$$
$$q: \text{factors of } 2 = \pm 1, \pm 2$$

The possibilities of $\frac{p}{q}$, in simplest form, are ± 1, $\pm\frac{1}{2}$, ± 3, and $\pm\frac{3}{2}$.

These values can be tested by using direct substitution or by using synthetic division and finding the remainder. Synthetic division is the more efficient method because if a zero is found, the polynomial can be written in factored form and, if possible, can be factored further using more traditional methods.

Example 2: Find rational zeros of $f(x) = 2x^3 + 3x^2 - 8x + 3$ by using synthetic division.

Each line in the following table represents the "quotient line" or the "answer line," when synthetic division is used. The column under "3" represents the remainder in synthetic division. ***Remember:*** When there is a remainder, we do *not* have a factor.

$\frac{p}{q}$	2	3	–8	3	
1	2	5	–3	0	1 is a zero.
–1	2	1	–9	12	
$\frac{1}{2}$	2	4	–6	0	$\frac{1}{2}$ is a zero.
$-\frac{1}{2}$	2	2	–9	$\frac{15}{2}$	
3	2	9	19	60	
–3	2	–3	1	0	–3 is a zero.
$\frac{3}{2}$	2	6	1	$\frac{9}{2}$	
$-\frac{3}{2}$	2	0	–8	15	

The zeros of $f(x) = 2x^3 + 3x^2 - 8x + 3$ are 1, $\frac{1}{2}$, and -3. This means

$$f(1)=0,\ f\left(\frac{1}{2}\right)=0,\ \text{and}\ f(-3)=0$$

The zeros could have been found without doing so much synthetic division. From the first line of the chart, 1 is seen to be a zero. This allows $f(x)$ to be written in factored form using the synthetic division result.

$$f(x) = 2x^3 + 3x^2 - 8x + 3 = (x - 1)(2x^2 + 5x - 3)$$

But $2x^2 + 5x - 3$ can be further factored into $(2x - 1)(x + 3)$ using the more traditional methods of factoring.

$$2x^2 - 5x - 3 = (x - 1)(2x - 1)(x + 3)$$

From this completely factored form, the zeros are quickly recognized. Zeros will occur when

$$x-1=0 \qquad 2x-1=0 \qquad x+3=0$$

$$x=1 \qquad x=\frac{1}{2} \qquad x=-3$$

Graphing Polynomial Functions

Polynomial functions of the form $f(x) = x^n$, where n is a positive integer, form one of two basic graphs:

$f(x)=x^n$, n an even integer

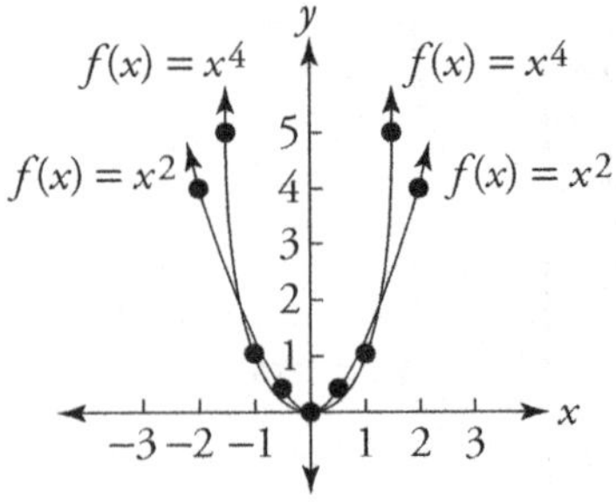

(a)

$f(x)=x^n$, n an odd integer

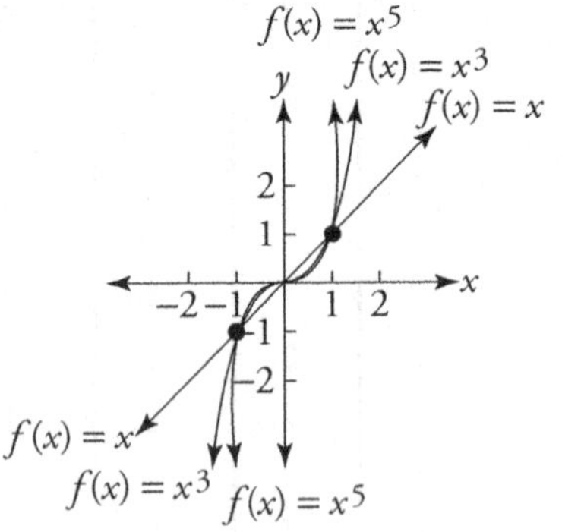

(b)

Each graph has the origin as its only x-intercept and y-intercept. Each graph contains the ordered pair (1, 1). If a polynomial function can be factored, its x-intercepts can be immediately found. Then a study is made as to what happens between these intercepts, to the left of the far left intercept and to the right of the far right intercept.

As shown in the graphs, when the leading term has an even exponent, both "ends" of the graph will point in the same direction. When the leading term has an odd exponent, the "ends" of the graph will point in opposite directions.

Example 1: Graph $f(x) = x^4 - 10x^2 + 9$.

$$\begin{aligned} f(x) &= x^4 - 10x^2 + 9 \text{ can be factored} \\ &= x^4 - 10x^2 + 9 \\ &= (x^2 - 1)(x^2 - 9) \\ &= (x+1)(x-1)(x+3)(x-3) \end{aligned}$$

The zeros of this function are -1, 1, -3, and 3. That is, -1, 1, -3, and 3 are the x-intercepts of this function.

The following process determines where the graph is in the intervals between the intercepts.

When $x < -3$, let's choose $x = -4$, then

$$\begin{aligned} f(x) &= (x+1)(x-1)(x+3)(x-3) \\ f(-4) &= (-4+1)(-4-1)(-4+3)(-4-3) \\ &= (-3)(-5)(-1)(-7) \\ &= 105 \end{aligned}$$

So for $x < -3$, $f(x) > 0$.

When $-1 < x < 1$, let's choose $x = 0$; then

$$\begin{aligned} f(x) &= (x+1)(x-1)(x+3)(x-3) \\ f(0) &= (0+1)(0-1)(0+3)(0-3) \\ &= (1)(-1)(3)(-3) \\ &= 9 \end{aligned}$$

So for $-1 < x < 1$, $f(x) > 0$.

In a similar way, it can be seen that

when $x > 3, f(x) > 0$,
when $-3 < x < -1, f(x) < 0$, and
when $1 < x < 3, f(x) < 0$.

The graph then has points in the shaded regions, as shown in the following graph:

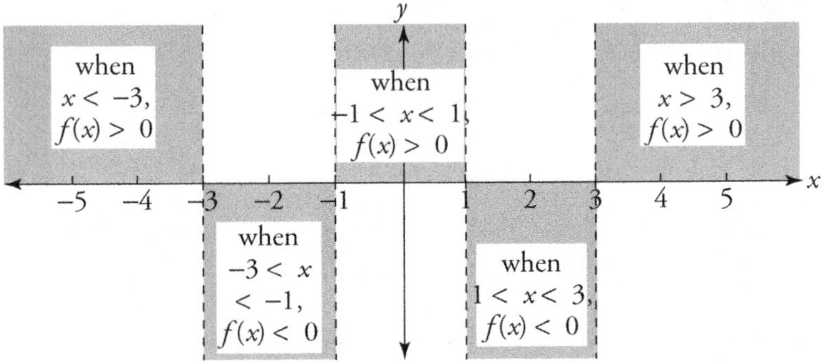

The y-intercept of this function is found by finding $f(0)$.

$f(0) = 9$, so $(0, 9)$ is a point on the graph. To complete the graph, find and plot several points. Evaluate $f(x)$ for several integer replacements, then connect those points to form a smooth curve, as shown here:

x	$f(x)$
4	105
3	0
2	15
1	0
0	9
1	0
2	15
3	0
4	150

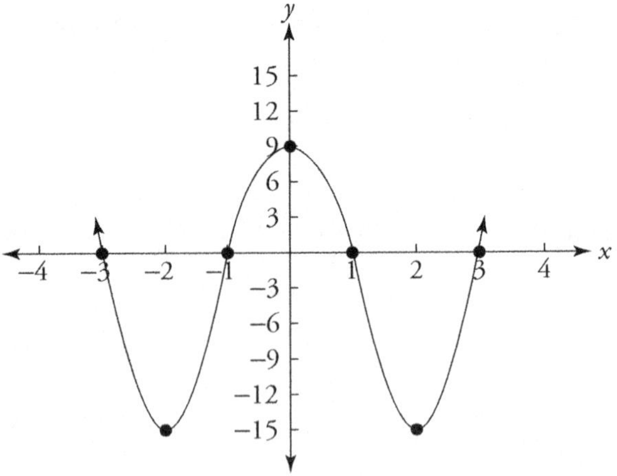

Notice that $f(x) = x^4 - 10x^2 + 9$ has a leading term with an even exponent. The far right and far left sides of the graph will point in the same direction. Because the leading coefficient is positive, the two sides will go up. If the leading coefficient were negative, the two sides would go down.

Example 2: Graph $f(x) = x^3 - 19x + 30$.

$f(x) = x^3 - 19x + 30$ can be factored using the rational zero theorem:

$\frac{p}{q}$	1	0	–19	30
1	1	1	–18	12
–1	1	–1	–18	48
2	1	2	–15	0

$f(x)$ can now be written in factored form: $f(x) = (x - 2)(x^2 + 2x - 15)$. And further factored to $f(x) = (x - 2)(x - 3)(x + 5)$.

The zeros of this function are 2, 3, and -5 (see the table and graph that follow).

The x-intercepts are (2, 0), (3, 0), and $(-5, 0)$.

x	$f(x)$
–6	–72
–5	0
–4	42
–3	60
–2	60
–1	48
0	30
1	12
2	0
3	0
4	18
5	60

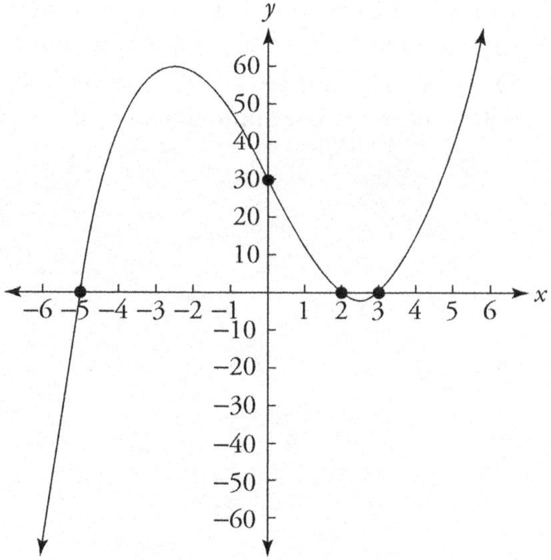

Notice that $f(x) = x^3 - 19x + 30$ has a leading term that has a positive coefficient and an odd exponent. This function will always go up toward the far right and down toward the far left. If the leading coefficient were negative with an odd exponent, the graph would go up toward the far left and down toward the far right.

Chapter Check-Out

Questions

1. Find $P(-2)$ if $P(x) = 6x^4 - 5x^3 + 2x - 12$.
2. Find the zeros of the following function: $f(x) = x^2 - x - 12$.
3. Find all rational zeros of the following function: $f(x) = 2x^3 + 7x^2 + 2x - 3$.
4. Find the zeros of the following function: $f(x) = x^3 + 4x^2 + x - 6$.
5. Explain the relationship between the following terms:
   (a) Degree of the polynomial
   (b) Zeros of the function
   (c) x-intercepts of the graph of the function
6. Explain the steps used to sketch the graph of a polynomial.
7. Explain the effect that the coefficient and the exponent of the leading term have on the graph of a polynomial.
8. Name the two different methods that can be used to find the remainder when one polynomial is divided by another.
9. The zeros of a polynomial function are the_______-coordinate of the ______-intercept(s) of the graph of the function.
10. Describe the graph for $f(x) = x^3 - 3x^2 + 3$. (***Note:*** Not all intercepts will be rational. Use approximation for irrational intercepts.)

Answers

1. 120
2. $-3, 4$
3. $-3, -1, \frac{1}{2}$
4. $1, -2, -3$
5. (a) highest exponent; The degree of a polynomial is the sum of the exponents of the variables that have the greatest sum.

   (b) The zeros of a function are any replacements for the variable that provide an answer of zero.

   (c) where the graph crosses the x-axis, or x-intercepts; The x-intercepts of the graph of a function are the zeros of a function.
6. Find the possible roots and use synthetic division. Depending on the exponent of the leading term, the graph at both ends will either point in the same direction or point in the opposite direction. Then plot the x-intercepts. Next, pick points in the intervals. Finally, sketch the graph (see Examples 1 and 2 on pp. 171–174).
7. The coefficient affects the up/down direction. The exponent affects whether both "ends" of the graph go in the same direction or in the opposite direction. The factor of the coefficient of the leading term (q) and the factor of the constant term (p) form $\pm\frac{p}{q}$. This ratio is used to find the possible rational zeros.
8. Synthetic division or direct substitution of the constant term
9. x-coordinate; x-intercept(s)
10. Choose at least three values of the x-coordinates, and substitute into the original equation to find the corresponding y-coordinate of other points on the graph. Left "end" points down; right "end" points up. Contains points such as $(-1, -1)$, $(0, 3)$, $(1, 1)$, $(2, -1)$, and $(3, 3)$. There are three x-intercepts. The y-intercept is at $(0, 3)$.

Chapter 10

RADICALS AND COMPLEX NUMBERS

Chapter Check-In

- ❑ Defining and simplifying radicals
- ❑ Working with radical expressions: Adding, subtracting, multiplying, and dividing
- ❑ Defining and simplifying rational exponents
- ❑ Defining and working with complex numbers

Common Core Standard: The Real and Complex Number System and Seeing the Structure in Expressions

Extend the properties of exponents to rational exponents (N.RN.1-2). Use complex numbers in polynomial identities and equations (N.RN. 7-9). Reason with equations and inequalities (A.REI.1, 2, 4, 6, 7, 11).

Radicals are the "undoing" of exponents. In other words, since 3 squared is 9, radical 9 is 3. Symbolically, $3^2 = 9$; $\sqrt{9} = 3$. The radical sign, $\sqrt{\ }$, is used to indicate "the root" of the number beneath it. If the radical sign has no index, which is the number written in its leading crook (like this $\sqrt[3]{\ }$, indicating cube root), then it implies the square root of the number under the sign that is being indicated. This chapter is concerned with the terminology and operations involving radical expressions. When performing arithmetic operations on radicals, the radical term behaves as if it were a variable.

Complex numbers are natural partners to radicals. They result from finding the square root of a negative expression. The imaginary number, i, is defined as the square root of -1. $\sqrt{-9}$ can be simplified to $\left(\sqrt{9}\right)\left(\sqrt{-1}\right) = 3i$. Complex numbers are addressed starting on p. 187.

Radicals

The expression $\sqrt[n]{a}$ is called a **radical expression.** The symbol $\sqrt{\ }$ is called the **radical sign.** The expression under the radical sign is called the **radicand,** and n, an integer greater than 1, is called the **index.** If the radical expression appears without an index, the index is assumed to be 2. The expression $\sqrt[n]{a}$ is read as "the nth root of a." By definition:

$$x=\sqrt[n]{a}, \text{ then } x^n = a$$

Example 1:

(a) Simplify $\sqrt{25}$.

When $x=\sqrt{25}$, then $x^2 = 25$. So, $x = 5$ or $x = -5$.

Because x could be either value, a rule is established. Since a numerical radical expression could have either a positive or a negative answer, always take the positive. This is called the *principal root.* Thus, $\sqrt{25}=5$ and $-\sqrt{25}=-5$.

(b) Simplify $\sqrt[3]{64}$.

When $x=\sqrt[3]{64}$, then $x^3 = 64$ and $x = 4$. So, $\sqrt[3]{64}=4$.

(c) Simplify $\sqrt[5]{\frac{1}{32}}$.

When $x=\sqrt[5]{\frac{1}{32}}$, then $x^5=\frac{1}{32}$ and $x=\frac{1}{2}$. So, $\sqrt[5]{\frac{1}{32}}=\frac{1}{2}$.

(d) Simplify $\sqrt[3]{-\frac{1}{8}}$.

When $x=\sqrt[3]{-\frac{1}{8}}$, then $x^3=-\frac{1}{8}$ and $x=-\frac{1}{2}$. So, $\sqrt[3]{-\frac{1}{8}}=-\frac{1}{2}$.

(e) Simplify $\sqrt{-4}$.

When $x=\sqrt{-4}$, then $x^2 = -4$. There is no *real* value for x, so $\sqrt{-4}$ is *not a real number.*

Following are true statements regarding radical expressions.

n	*a*	$\sqrt[n]{a}$	Example
even	positive	positive	$\sqrt[4]{16}=2$
	negative	not real	$\sqrt[4]{-16}$ not real
	zero	zero	$\sqrt[4]{0}=0$
odd	positive	positive	$\sqrt[3]{8}=2$
	negative	negative	$\sqrt[3]{-8}=-2$
	zero	zero	$\sqrt[3]{0}=0$

Note: When the index is *even,* the radicand must be nonnegative for the answer to be real. When the index is *odd,* there are no restrictions on the radicand.

When variables are involved, absolute value signs are often needed.

Example 2: Simplify $\sqrt{x^2}$.

It would seem that $\sqrt{x^2}=x$. But there is no guarantee that x is nonnegative. Because of this, $\sqrt{x^2}=|x|$. This guarantees that the result is nonnegative.

Absolute value signs are *never* used when the index is odd. Absolute value signs are often used when the index is even, and the result could possibly be negative.

Example 3: Simplify the following, using absolute value signs when needed.

(a) Simplify $\sqrt[4]{16x^8y^{24}}$.

$$\sqrt[4]{16x^8y^{24}}=\sqrt[4]{2^4(x^2)^4(y^6)^4}=\left|2x^2y^6\right|=2x^2y^6$$

Since $2x^2y^6$ could not be negative even if x or y were negative, absolute value signs are not needed.

(b) Simplify $\sqrt[4]{16x^8y^{20}}$.

$$\sqrt[4]{16x^8y^{20}} = \sqrt[4]{2^4(x^2)^4(y^5)^4} = |2x^2y^5|$$

Since the expression could be negative if y were negative, a correct way to represent the answer is $|2x^2y^5|$. Since only y could have caused the answer to be negative, another way to represent the answer is $2x^2|y|^5$.

(c) Simplify $\sqrt[5]{32x^{10}y^{15}}$.

$$\sqrt[5]{32x^{10}y^{15}} = \sqrt[5]{2^5(x^2)^5(y^3)^5} = 2x^2y^3$$

Absolute value signs are *never* used when the index is odd.

Simplifying Radicals

Simplifying radicals involves the product rule.

> **Product Rule for Radicals**
> If n is even, and $a \geq 0$ and $b \geq 0$, then $\sqrt[n]{ab} = \sqrt[n]{a}\sqrt[n]{b}$.
> If n is odd, then for all real numbers a and b, $\sqrt[n]{ab} = \sqrt[n]{a}\sqrt[n]{b}$.

Remember: Even roots—the radicand must be nonnegative. Odd roots—there is no restriction on the radicand.

Example 1:

(a) Simplify $\sqrt{20}$.

$$\sqrt{20} = \sqrt{2^2 \cdot 5} = \sqrt{2^2} \cdot \sqrt{5} = 2\sqrt{5}$$

(b) Simplify $\sqrt[3]{64x^9y^{10}z^4}$.

$$\sqrt[3]{64x^9y^{10}z^4} = \sqrt[3]{(4)^3(x^3)^3(y^3)^3z^3yz}$$

$$= \sqrt[3]{(4)^3(x^3)^3(y^3)^3z^3} \Diamond \sqrt[3]{yz} = 4x^3y^3z^3\sqrt{yz}$$

(c) Simplify $\sqrt{(x-3)^6}$.

$$\sqrt{(x-3)^6} = \sqrt{\left[(x-3)^3\right]^2} = |(x-3)^3| \text{ or } |x-3|^3$$

(d) Simplify $\sqrt[3]{(y-5)^5}$.

$$\sqrt[3]{(y-5)^5}=\sqrt[3]{(y-5)^3(y-5)^2}=\sqrt[3]{(y-5)^3}\cdot\sqrt[3]{(y-5)^2}$$
$$=(y-5)\sqrt[3]{(y-5)^2}$$

Example 2: Simplify $\sqrt[3]{12x^2}\cdot\sqrt[3]{4x^4}$.

$$\sqrt[3]{12x^2}\cdot\sqrt[3]{4x^4}=\sqrt[3]{(12x^2)(4x^4)}$$
$$=\sqrt[3]{48x^6}$$
$$=\sqrt[3]{2^3\cdot 2\cdot 3\left(x^2\right)^3}$$
$$=\sqrt[3]{2^3\cdot 6\left(x^2\right)^3}$$
$$=2x^2\sqrt[3]{6}$$

Adding and Subtracting Radical Expressions

Radical expressions are called **like radical expressions** if the indexes are the same and the radicands are identical. For example, $\sqrt[3]{5x}$ and $4\sqrt[3]{5x}$ are like radical expressions because the indexes are the same and the radicands are identical. However, $\sqrt[3]{5x}$ and $\sqrt[3]{5y}$ are not like radical expressions because their radicands are not identical. Radical expressions can be added or subtracted only if they are *like radical expressions.*

Example 1:

(a) Simplify $4+3\sqrt{7}+6+5\sqrt{7}$.

$$4+3\sqrt{7}+6+5\sqrt{7}=(4+6)+\left(3\sqrt{7}+5\sqrt{7}\right)$$
$$=10+8\sqrt{7}$$

(b) Simplify $5\sqrt{27}+6\sqrt{3}-4\sqrt{48}$.

$$\begin{aligned}5\sqrt{27}+6\sqrt{3}-4\sqrt{48} &= 5\sqrt{9\cdot 3}+6\sqrt{3}-4\sqrt{16\cdot 3}\\ &= 5\sqrt{3^2\cdot 3}+6\sqrt{3}-4\sqrt{4^2\cdot 3}\\ &= 5\cdot 3\sqrt{3}+6\sqrt{3}-4\cdot 4\sqrt{3}\\ &= 15\sqrt{3}+6\sqrt{3}-16\sqrt{3}\\ &= 5\sqrt{3}\end{aligned}$$

(c) Simplify $\sqrt[3]{24a}+\sqrt[3]{81a}$.

$$\begin{aligned}\sqrt[3]{24a}+\sqrt[3]{81a} &= \sqrt[3]{8\cdot 3a}+\sqrt[3]{27\cdot 3a}\\ &= \sqrt[3]{2^3\cdot 3a}+\sqrt[3]{3^3\cdot 3a}\\ &= 2\sqrt[3]{3a}+3\sqrt[3]{3a}\\ &= 5\sqrt[3]{3a}\end{aligned}$$

Multiplying Radical Expressions

To multiply radical expressions, use the distributive property and the product rule for radicals.

Example 1:

(a) Simplify $(4+3\sqrt{2})(\sqrt{10}+\sqrt{5})$.

Use the F.O.I.L. method.

$$\begin{aligned}\left(4+3\sqrt{2}\right)\left(\sqrt{10}+\sqrt{5}\right) &= 4\sqrt{10}+4\sqrt{5}+3\sqrt{2}\cdot\sqrt{10}+3\sqrt{2}\cdot\sqrt{5}\\ &= 4\sqrt{10}+4\sqrt{5}+3\sqrt{20}+3\sqrt{10}\\ &= 4\sqrt{10}+4\sqrt{5}+3\sqrt{4\cdot 5}+3\sqrt{10}\\ &= 4\sqrt{10}+4\sqrt{5}+6\sqrt{5}+3\sqrt{10}\\ &= 7\sqrt{10}+10\sqrt{5}\end{aligned}$$

(b) Simplify $\sqrt{6}(\sqrt{3}+\sqrt{12})$.

Use the distributive property.

$$\begin{aligned}\sqrt{6}\left(\sqrt{3}+\sqrt{12}\right) &= \sqrt{6}\cdot\sqrt{3}+\sqrt{6}\cdot\sqrt{12}\\ &= \sqrt{18}+\sqrt{72}\\ &= \sqrt{9\cdot 2}+\sqrt{36\cdot 2}\\ &= 3\sqrt{2}+6\sqrt{2}\\ &= 9\sqrt{2}\end{aligned}$$

Dividing Radical Expressions

When dividing radical expressions, use the quotient rule.

Quotient Rule for Radicals

For all real values, a and b,

1. if n is even, $a \geq 0$, and $b > 0$, then $\sqrt[n]{\frac{a}{b}} = \frac{\sqrt[n]{a}}{\sqrt[n]{b}}$, and
2. if n is odd and $b \neq 0$, then $\sqrt[n]{\frac{a}{b}} = \frac{\sqrt[n]{a}}{\sqrt[n]{b}}$.

Remember: Even roots—the radicand must be nonnegative. Odd roots—there is no restriction on the radicand. Denominators can never equal zero.

Radical expressions are written in simplest terms when

- The index is as small as possible.
- The radicand contains no factor other than 1, which is the nth or greater power of an integer or polynomial.
- The radicand contains no fractions.
- No radicals appear in the denominator.

Example 1:

(a) Simplify $\sqrt[3]{\frac{3}{8}}$.

Using the quotient rule for radicals,

$$\sqrt[3]{\frac{3}{8}} = \frac{\sqrt[3]{3}}{\sqrt[3]{2 \cdot 2 \cdot 2}}$$
$$= \frac{\sqrt[3]{3}}{2}$$

(b) Simplify $\frac{4\sqrt{14}}{8\sqrt{2}}$.

Using the quotient rule for radicals,

$$\frac{4\sqrt{14}}{8\sqrt{2}} = \frac{4}{8} \cdot \frac{\sqrt{14}}{\sqrt{2}}$$
$$= \frac{1}{2} \cdot \sqrt{\frac{14}{2}}$$
$$= \frac{1}{2} \cdot \sqrt{7}$$
$$= \frac{\sqrt{7}}{2}$$

Rationalizing the denominator

An expression with a radical in its denominator needs to be simplified into one *without* a radical in its denominator. This process is called **rationalizing the denominator.** This is accomplished by multiplying the expression by a fraction whose value = 1.

Example 2:

(a) Simplify $\sqrt{\frac{7}{x}}$.

To rationalize this denominator, the fraction whose value = 1 is $\frac{\sqrt{x}}{\sqrt{x}}$; that will eliminate the radical in the denominator.

$$\sqrt{\frac{7}{x}} = \frac{\sqrt{7}}{\sqrt{x}} \Diamond \frac{\sqrt{x}}{\sqrt{x}}$$
$$= \frac{\sqrt{7x}}{\sqrt{x^2}}$$
$$= \frac{\sqrt{7x}}{x}$$

Note: We elected to use the principal root of $\sqrt{x^2}$. Absolute value isn't needed here; it's assumed that x is positive because it had to be in order to write the initial expression.

(b) Simplify $\sqrt[3]{\frac{12x^2}{5y^2}}$.

$$\sqrt[3]{\frac{12x^2}{5y^2}} = \frac{\sqrt[3]{12x^2}}{\sqrt[3]{5y^2}}$$

To rationalize the denominator, ask: "What can $\sqrt[3]{5y^2}$ be multiplied by so that the result will not involve a radical?" The answer is $\sqrt[3]{5^2 y}$ or $\sqrt[3]{25y}$. That choice is made so that after they are multiplied, everything in the denominator under the radical sign will be perfect cubes.

$$\sqrt[3]{\frac{12x^2}{5y^2}} = \frac{\sqrt[3]{12x^2}}{\sqrt[3]{5y^2}} \cdot \frac{\sqrt[3]{25y}}{\sqrt[3]{25y}}$$
$$= \frac{\sqrt[3]{300x^2 y}}{\sqrt[3]{125y^3}}$$
$$= \frac{\sqrt[3]{300x^2 y}}{\sqrt[3]{5y \cdot 5y \cdot 5y}}$$
$$= \frac{\sqrt[3]{300x^2 y}}{5y}$$

Conjugates

If a and b are unlike terms, then the **conjugate** of $a + b$ is $a - b$, and the conjugate of $a - b$ is $a + b$. The conjugate of $(5+\sqrt{3})$ is $(5-\sqrt{3})$. Conjugates are used for rationalizing the denominator when the denominator is a two-termed expression involving a square root.

Example 3: Simplify $\frac{6+\sqrt{3}}{4-\sqrt{3}}$.

To rationalize the denominator of this expression, multiply by a fraction whose value = 1 in the form of the denominator's conjugate. F.O.I.L. both the numerator and the denominator.

$$\begin{aligned}\frac{6+\sqrt{3}}{4-\sqrt{3}} &= \frac{6+\sqrt{3}}{4-\sqrt{3}} \cdot \frac{4+\sqrt{3}}{4+\sqrt{3}} \\ &= \frac{24+6\sqrt{3}+4\sqrt{3}+3}{16+4\sqrt{3}-4\sqrt{3}-3} \\ &= \frac{27+10\sqrt{3}}{13}\end{aligned}$$

Rational Exponents

If n is a natural number greater than 1 and b is any real number, then $b^{\frac{1}{n}} = \sqrt[n]{b}$. And since a negative exponent indicates a reciprocal, then $b^{-\frac{1}{n}} = \frac{1}{\sqrt[n]{b}}$ when $b \neq 0$.

Example 1:

(a) Simplify $36^{\frac{1}{2}}$.

$$36^{\frac{1}{2}} = \sqrt{36} = 6$$

(b) Simplify $(72x^4y)^{\frac{1}{3}}$.

$$(72x^4y)^{\frac{1}{3}} = \sqrt[3]{72x^4y} = \sqrt[3]{8 \cdot 9x^4y} = \sqrt[3]{2^3 \cdot 9 \cdot x^3 \cdot xy} = 2x\sqrt[3]{9xy}$$

(c) Simplify $16^{\left(-\frac{1}{4}\right)}$.

$$16^{\left(-\frac{1}{4}\right)} = \frac{1}{\sqrt[4]{16}} = \frac{1}{\sqrt[4]{2^4}} = \frac{1}{2}$$

Hence, the definition of a **rational exponent** is as follows: If n is a natural number greater than 1, m is an integer, and b is a nonnegative real number, then $b^{\frac{m}{n}} = \left(\sqrt[n]{b}\right)^m$. Note that $\left(\sqrt[n]{b}\right)^m$ can also be written as $\sqrt[n]{b^m}$.

Example 2:

(a) Simplify $27^{\frac{2}{3}}$.

$$27^{\frac{2}{3}} = \sqrt[3]{27^2} = \left(\sqrt[3]{27}\right)^2 = \left(\sqrt[3]{3^3}\right)^2 = (3)^2 = 9$$

(b) Simplify $25^{-\left(\frac{3}{2}\right)}$.

$$25^{-\left(\frac{3}{2}\right)} = \sqrt{25^{-3}} = \left(\sqrt{25}\right)^{-3} = 5^{-3} = \frac{1}{5^3} = \frac{1}{125}$$

Complex Numbers

The expression $\sqrt{-1}$ has no real answer. The symbol i is created to represent $\sqrt{-1}$ and is called an **imaginary number.** Since $i = \sqrt{-1}$, $i^2 = -1$. Any expression that is a product of a real number with i is called a **pure imaginary number.**

Example 1:

(a) Simplify $\sqrt{-25}$.

$$\sqrt{-25} = \sqrt{25} \cdot \sqrt{-1} = 5i$$

(b) Simplify $\sqrt{-40}$.

$$\sqrt{-40} = \sqrt{40} \cdot \sqrt{-1} = \sqrt{4 \cdot 10} \cdot \sqrt{-1} = 2\sqrt{10} \cdot \sqrt{-1} = 2\sqrt{10} \cdot i$$

This last expression is commonly written as $2i\sqrt{10}$ so that the i is not mistakenly written under the radical.

(c) Simplify $(6i)(4i)$.

$$(6i)(4i) = 24i^2 = 24(-1) = -24$$

(d) Simplify $\sqrt{-6} \cdot \sqrt{-8}$.

$$\sqrt{-6} \cdot \sqrt{-8} = \left(i\sqrt{6}\right)\left(i\sqrt{8}\right) = i^2\sqrt{48} = -1\left(4\sqrt{3}\right) = -4\sqrt{3}$$

For this last example, all imaginary values had to be put into their "*i*-form" *before* any simplifying could be done. Note that

$$\sqrt{-6} \cdot \sqrt{-8} \neq \sqrt{(-6)(-8)}$$
$$\left(i\sqrt{6}\right)\left(i\sqrt{8}\right) \neq \sqrt{48}$$
$$i^2\sqrt{48} \neq \sqrt{48}$$
$$-1\sqrt{48} \neq \sqrt{48}$$

That is, the product rule for radicals *does not hold* (in general) with imaginary numbers. Simplify first, then multiply.

When i is raised to powers, it has a repeating pattern.

$i^0 = 1$	$i^4 = 1$	$i^8 = 1$	$i^{12} = 1$
$i^1 = i$	$i^5 = i$	$i^9 = i$	$i^{13} = i$
$i^2 = -1$	$i^6 = -1$	$i^{10} = -1$	$i^{14} = -1$
$i^3 = -i$	$i^7 = -i$	$i^{11} = -i$	$i^{15} = -i$

When i is raised to any whole-number power, the result is always 1, i, -1, or $-i$. If the exponent on i is divided by 4, the remainder indicates which of the four values is the result. This reads "i runs in cycles of four."

Example 2:

(a) Simplify i^{34}.

Since 34 divided by 4 has a remainder of 2, $i^{34} = i^2 = -1$. Think of it as $i^{4k} = 1$; so, $i^{34} = i^{32} \cdot i^2 = 1 \cdot i^2 = 1(-1) = -1$.

(b) Simplify i^{95}.

Since 95 divided by 4 has a remainder of 3, $i^{95} = i^3 = -i$.

(c) Simplify i^{108}.

Since 108 divided by 4 has a zero remainder, $i^{108} = i^0 = 1$.

(d) Simplify i^{53}.

Since 53 divided by 4 has a remainder of 1, $i^{53} = i^1 = i$.

A **complex number** is any expression that is a sum of a pure imaginary number and a real number. A complex number usually is expressed in a form called the $a + bi$ form, or standard form, where a and b are real numbers. The expressions $a + bi$ and $a - bi$ are called **complex conjugates.** Complex conjugates are used to rationalize the denominator when dividing with complex numbers. It is equivalent to rationalizing the denominator when dealing with fractions. The objective is to clear the denominator of radicals and complex numbers.

Arithmetic with complex numbers is done in a similar manner as arithmetic with polynomials. The following are definitions for arithmetic with two complex numbers called $(a + bi)$ and $(c + di)$.

- Combining like terms and factoring out the i,

$$(a+bi)+(c+di)=(a+c)+(b+d)i$$
$$(a+bi)-(c+di)=(a-c)+(b-d)i$$

- Using the double distributive property, or F.O.I.L.,

$$\begin{aligned}(a+bi)(c+di) &= ac+adi+bci+bdi^2\\ &= ac+adi+bci-bd\\ &= (ac-bd)+(ad+bc)i\end{aligned}$$

  Reminder: $i^2 = -1$.

- Rationalizing the denominator,

$$\frac{a+bi}{c+di}=\underbrace{\frac{a+bi}{c+di}\cdot\frac{c-di}{c-di}}_{\text{simplifying the denominator}}=\frac{(ac+bd)+(bc-ad)i}{c^2+d^2}$$

Example 3: Find the sum, difference, product, and quotient of $(4 + 3i)$ and $(5 - 4i)$.

$$\text{Sum:}\quad (4+3i)+(5-4i)=(4+5)+(3-4)i$$
$$=9-i$$

$$\text{Difference:}\quad (4+3i)-(5-4i)=(4-5)+[3-(-4)]i$$
$$=-1+7i$$

$$\text{Product:}\quad (4+3i)(5-4i)=20-16i+15i-12i^2$$
$$=20-i+12$$
$$=32-i$$

Quotient: Rationalize the denominator.

$$\frac{4+3i}{5-4i}=\frac{4+3i}{5-4i}\cdot\frac{5+4i}{5+4i}$$
$$=\frac{(4\cdot 5)+(4\cdot 4i)+(5\cdot 3i)+[(3i)(4i)]}{25+20i-20i-16i^2}$$
$$=\frac{20+16i+15i+12i^2}{25-16(-1)}$$
$$=\frac{20+31i+12(-1)}{25+16}$$
$$=\frac{20-12+31i}{41}$$
$$=\frac{8+31i}{41}$$
$$=\frac{8}{41}+\frac{31}{41}i$$

Example 4: Simplify $\frac{5}{6i}$.

Since $6i$ can be considered $0 + 6i$, its complex conjugate is $0 - 6i$. Therefore,

$$\begin{aligned}\frac{5}{6i} &= \frac{5}{6i} \cdot \frac{-6i}{-6i} \\ &= \frac{-30i}{-36i^2} \\ &= \frac{-30i}{36} \\ &= -\frac{5}{6}i\end{aligned}$$

Chapter Check-Out

Questions

1. Simplify each of the following:
   (a) $\sqrt{81}$
   (b) $\sqrt[4]{256}$
   (c) $\sqrt[3]{216}$
   (d) $\sqrt{-16}$
2. Simplify each of the following:
   (a) $\sqrt{24}$
   (b) $\sqrt{162}$
   (c) $\sqrt[3]{8c^6d^8f^9}$
   (d) $27^{-\left(\frac{1}{3}\right)}$
3. Find the sum, difference, product, and quotient of $(6 + 3i)$ and $(4 - 2i)$ in that order. Rationalize any denominators if appropriate.
4. Explain the difference between a pure imaginary number and a complex number.
5. Why does it matter if the index is even or odd?
6. Simplify $\sqrt{\frac{1}{2}} + \sqrt{\frac{2}{3}} - \sqrt{\frac{9}{8}}$.

Answers

1. **(a)** 9; **(b)** 4; **(c)** 6; **(d)** $4i$
2. **(a)** $2\sqrt{6}$; **(b)** $9\sqrt{2}$; **(c)** $2c^2d^2f^3\sqrt[3]{d^2}$; **(d)** $\frac{1}{3}$
3. Sum: $10 + i$; difference: $2 + 5i$; product: 30; quotient: $\frac{9}{10}+\frac{6}{5}i$
4. A pure imaginary number is in the form of bi, where $b \neq 0$. For example, $7i$. A complex number is made up of a real number and an imaginary number. It is in the form of $a + bi$, where $a \neq 0$ and $b \neq 0$. For example, $5 + 7i$.
5. An odd index produces no restriction on the radicand.
6. $\frac{-\sqrt{2}}{4}+\frac{\sqrt{6}}{3}$

Chapter 11

QUADRATICS IN ONE VARIABLE

Chapter Check-In

- ❑ Solving quadratics by factoring
- ❑ Solving quadratics by the square root property
- ❑ Solving quadratics by completing the square
- ❑ Solving quadratics by the quadratic formula
- ❑ Solving equations in quadratic form
- ❑ Solving radical equations
- ❑ Solving quadratic inequalities

Common Core Standard: Linear, Quadratic, and Exponential Models

Solve quadratic equations and equations in quadratic form by any method, including graphing and graphing quadratic inequalities. Construct and compare linear, quadratic, and exponential models and solve problems. Observe, using graphs and tables, that a quantity increasing exponentially eventually exceeds a quantity increasing linearly, quadratically, or as a polynomial function (F.LE.3). Build new functions from existing functions (F.BF.3). Solve quadratic equations in one variable (A.REI.4). Understand the relationship between zeros of functions and factors of polynomials (A.APR.3).

The quadratic equation is arguably the signature entity of algebra. Everything you have learned up to this point will be used to solve quadratic equations and inequalities. This chapter takes you through the logical sequence of methods that can be used to solve quadratics.

Quadratic Equations

A **quadratic equation** is any equation of the form

$$ax^2 + bx + c = 0 \qquad (a \neq 0)$$

A quadratic equation is usually solved in one of four algebraic ways:

- Factoring
- Applying the square root property
- Completing the square
- Using the quadratic formula

Solving Quadratics by Factoring

Solving a quadratic equation by factoring depends on the zero-product rule. The **zero-product rule** states that if $ab = 0$, then either $a = 0$, $b = 0$, or both.

Example 1: Solve $2x^2 = -9x - 4$ for x by factoring.

First, get all terms on one side of the equation.

$$2x^2 = -9x - 4$$

$$2x^2 + 9x + 4 = 0$$

Factor the quadratic.

$$(2x + 1)(x + 4) = 0$$

Apply the zero-product rule.

$$2x + 1 = 0 \quad \text{or} \quad x + 4 = 0$$

$$x = -\frac{1}{2} \qquad\qquad x = -4$$

Solving Quadratics by the Square Root Property

The **square root property** says that if $x^2 = c$, then $x = \sqrt{c}$ or $x = -\sqrt{c}$. This can be written as "if $x^2 = c$, then $x = \pm\sqrt{c}$." If c is positive, then x has two real answers. If c is negative, then x has two imaginary answers.

Example 1: Solve each of the following equations for x using the square root property.

(a) $x^2 = 48$

$$x = \pm\sqrt{48}$$
$$= \pm 4\sqrt{3}$$

(b) $x^2 = -16$

$$x = \pm\sqrt{-16}$$
$$= \pm 4i$$

(c) $5x^2 - 45 = 0$

$$5x^2 = 45$$
$$x^2 = 9$$
$$x = \pm\sqrt{9}$$
$$= \pm 3$$

(d) $(x-7)^2 = 81$

$$(x-7) = \pm\sqrt{81}$$
$$x - 7 = \pm 9$$
$$x = 7 \pm 9$$
$$x = 16 \text{ or } -2$$

(e) $(x+3)^2 = 24$

$$(x+3) = \pm\sqrt{24}$$
$$x + 3 = \pm 2\sqrt{6}$$
$$x = -3 \pm 2\sqrt{6}$$
$$x = -3 + 2\sqrt{6} \text{ or } x = -3 - 2\sqrt{6}$$

Solving Quadratics by Completing the Square

The expression $x^2 + bx$ can be made into a square trinomial by adding a certain value to it. This value is found by performing two steps:

1. Multiply b (the coefficient of the "x term") by $\frac{1}{2}$.
2. Square the result.

Example 1: Find the value to add to $x^2 + 8x$ to make it become a square trinomial.

$$x^2 + 8x$$

Multiply the coefficient of the "x term" by $\frac{1}{2}$.

$$8\left(\frac{1}{2}\right) = 4$$

Square that result.

$$(4)^2 = 16$$

So 16 must be added to $x^2 + 8x$ to make it a square trinomial.

$$x^2 + 8x + 16 = (x + 4)^2$$

Finding the value that makes a quadratic become a square trinomial is called **completing the square.** That square trinomial produces an equation that can then be solved by using the square root property.

Example 2: Solve the equation $x^2 - 10x = -16$ for x by using the completing the square method.

$$x^2 - 10x = -16$$

Multiply coefficient of "x term" by $\frac{1}{2}$.

$$(-10)\left(\frac{1}{2}\right) = -5$$

Square the result.

$$(-5)^2 = 25$$

Add 25 to both sides of the equation.

$$x^2 - 10x + 25 = -16 + 25$$

$$(x-5)^2 = 9$$

$$x - 5 = \pm\sqrt{9}$$

$$x - 5 = \pm 3$$

$$x = 5 \pm 3$$

$$x = 8 \text{ or } x = 2$$

To solve quadratic equations by using the completing the square method, the coefficient of the squared term must be 1. If it is not, then first divide all terms of the equation by that coefficient and then proceed as before.

Example 3: Solve $2x^2 - 3x + 4 = 0$ for x (solve by the method of completing the square).

Get the coefficient of the squared term to be 1.

$$2x^2 - 3x + 4 = 0$$

$$\frac{2x^2}{2} - \frac{3x}{2} + \frac{4}{2} = \frac{0}{2}$$

$$x^2 - \frac{3}{2}x + 2 = 0$$

Isolate the variable terms.

$$x^2 - \frac{3}{2}x = -2$$

Multiply coefficient of "x term" by $\frac{1}{2}$.

$$\left(-\frac{3}{2}\right)\left(\frac{1}{2}\right) = -\frac{3}{4}$$

Square that result.

$$\left(-\frac{3}{4}\right)^2 = \frac{9}{16}$$

The above process of completing the square multiplies the coefficient of x by $\frac{1}{2}$, squares the result, and adds that to both sides of the equation. Therefore, complete the square by adding $\frac{9}{16}$ to both sides of the equation. This process produces the following:

$$x^2 - \frac{3}{2}x + \frac{9}{16} = -2 + \frac{9}{16}$$

$$\left(x - \frac{3}{4}\right)^2 = -\frac{32}{16} + \frac{9}{16}$$

$$\left(x - \frac{3}{4}\right)^2 = -\frac{23}{16}$$

Use the square root property.

$$x-\frac{3}{4}=\pm\sqrt{-\frac{23}{16}}$$

$$x-\frac{3}{4}=\frac{\pm i\sqrt{-23}}{4}$$

$$x=\frac{3}{4}\pm\frac{i\sqrt{23}}{4}$$

$$x=\frac{3\pm i\sqrt{23}}{4}$$

Solving Quadratics by Formula

The following represents any quadratic equation:

$$ax^2 + bx + c = 0$$

This quadratic can be solved using the completing the square method to produce a formula that can then be applied to all quadratic equations. Following is the derivation of the quadratic formula.

Example 1: Solve $ax^2 + bx + c = 0$, $a \neq 0$, for x by using the completing the square method.

$$ax^2 + bx + c = 0$$

Get the coefficient of the squared term to be 1.

$$\frac{ax^2}{a}+\frac{bx}{a}+\frac{c}{a}=0$$

Isolate the variable terms.

$$x^2+\frac{b}{a}x+\frac{c}{a}=0$$

$$x^2+\frac{b}{a}x=-\frac{c}{a}$$

This is the process that completes the square.

$$\left(\frac{b}{a}\right)\left(\frac{1}{2}\right)=\frac{b}{2a}$$

$$\left(\frac{b}{2a}\right)^2=\frac{b^2}{4a^2}$$

Added to both sides to produce.

$$x^2+\frac{b}{a}x+\frac{b^2}{4a^2}=\frac{b^2}{4a^2}-\frac{c}{a}$$
$$\left(x+\frac{b}{2a}\right)^2=\frac{b^2}{4a^2}-\frac{4ac}{4a^2}$$
$$\left(x+\frac{b}{2a}\right)^2=\frac{b^2-4ac}{4a^2}$$

Apply the square root property.

$$x+\frac{b}{2a}=\pm\sqrt{\frac{b^2-4ac}{4a^2}}$$
$$x+\frac{b}{2a}=\frac{\pm\sqrt{b^2-4ac}}{2a}$$
$$x=\frac{-b\pm\sqrt{b^2-4ac}}{2a}$$

This end result, $x=\frac{-b\pm\sqrt{b^2-4ac}}{2a}$, is referred to as the **quadratic formula,** which can be used to solve *all* quadratic equations.

Quadratic Formula

$$x=\frac{-b\pm\sqrt{b^2-4ac}}{2a}$$

Example 2: Solve $2x^2 - 3x + 4 = 0$ for x by applying the quadratic formula.

$$2x^2-3x+4=0$$
$$a=2,\quad b=-3,\quad c=4$$

$$
\begin{aligned}
x &= \frac{-b \pm \sqrt{b^2 - 4ac}}{2a} \\
&= \frac{-(-3) \pm \sqrt{(-3)^2 - 4(2)(4)}}{2(2)} \\
&= \frac{3 \pm \sqrt{9 - 32}}{4} \\
&= \frac{3 \pm \sqrt{-23}}{4} \\
&= \frac{3 \pm i\sqrt{23}}{4}
\end{aligned}
$$

Note that this is the same problem solved in Example 3 on pp. 197–198 by completing the square. Here, however, it is solved by the quadratic formula.

The Discriminant

The expression under the radical sign is called the **discriminant.** It is used to tell the nature of the solutions, or roots, to the equation. It is a valuable tool in graphing quadratics, as will be shown in Chapter 12, "Conic Sections."

Once the values for a, b, and c are substituted, there are three possibilities:

$$b^2 - 4ac > 0, \qquad b^2 - 4ac = 0, \quad \text{or} \quad b^2 - 4ac < 0$$

When the discriminant is positive, there will be two real roots.

When the discriminant is equal to zero, there will be one real root.

When the discriminant is negative, there will be two imaginary roots.

The effect of this on the graph of the quadratic will be discussed again in Chapter 12.

Solving Equations in Quadratic Form

Any equation in the form $ax^{2n} + bx^{n} + c = 0$ is said to be in *quadratic form.* This equation then can be solved by using the quadratic formula, by completing the square, or by factoring.

Example 1: Solve $x^4 - 13x^2 + 36 = 0$ for x by **(a)** factoring and by **(b)** using the quadratic formula.

(a)

$$x^4 - 13x^2 + 36 = 0$$
$$(x^2 - 4)(x^2 - 9) = 0$$
$$(x+2)(x-2)(x+3)(x-3) = 0$$

By the zero-product rule,

$$x+2=0, \quad x-2=0, \quad x+3=0, \quad x-3=0$$
$$x=-2 \text{ or } x=2 \text{ or } x=-3 \text{ or } x=3$$

(b) $x^4 - 13x^2 + 36 = 0$ is equivalent to

$$\left(x^2\right)^2 - 13\left(x^2\right) + 36 = 0$$
$$a=1, \quad b=-13, \quad c=36$$

When applying the quadratic formula to equations in quadratic form, you are solving for the variable in the middle term. Thus, in this case,

$$x^2 = \frac{-(-13) \pm \sqrt{(-13)^2 - 4(1)(36)}}{2(1)}$$
$$= \frac{13 \pm \sqrt{169 - 144}}{2}$$
$$= \frac{13 \pm \sqrt{25}}{2}$$
$$= \frac{13 \pm 5}{2}$$
$$x^2 = 9 \text{ or } x^2 = 4$$

Using the square root property,

$$x = \pm\sqrt{9} \text{ or } x = \pm\sqrt{4}$$
$$= \pm 3 \qquad = \pm 2$$

Example 2: Solve $x-5\sqrt{x}-6=0$ for x by **(a)** factoring and by **(b)** using the quadratic formula.

(a)
$$x-5\sqrt{x}-6=0$$
$$\left(\sqrt{x}-6\right)\left(\sqrt{x}+1\right)=0$$
$$\sqrt{x}-6=0 \text{ or } \sqrt{x}+1=0$$
$$\sqrt{x}=6 \qquad \sqrt{x}=-1$$
$$\left(\sqrt{x}\right)^2=6^2$$
$$x=36$$

In the last step on the right, $\sqrt{x}$ must be a nonnegative value; therefore, $\sqrt{x}=-1$ has no solutions. The only solution is x = 36.

(b) $x-5\sqrt{x}-6=0$ is equivalent to
$$\left(\sqrt{x}\right)^2-5\left(\sqrt{x}\right)-6=0$$
$$a=1, \quad b=-5, \quad c=-6$$

When applying the quadratic formula to this quadratic form equation, you are solving for $\sqrt{x}$.

$$\sqrt{x}=\frac{-(-5)\pm\sqrt{(-5)^2-4(1)(-6)}}{2(1)}$$
$$=\frac{5\pm\sqrt{25+24}}{2}$$
$$=\frac{5\pm\sqrt{49}}{2}$$
$$=\frac{5\pm7}{2}$$
$$\sqrt{x}=6 \quad \text{or} \quad \sqrt{x}=-1$$
$$x=36$$

There is no real solution for $\sqrt{x}=-1$. Thus, x = 36 is the only solution.

Solving Radical Equations

A *radical equation* is an equation in which a variable is under a radical. This method is used if you don't recognize the equation as being in quadratic form. To solve a radical equation, follow these steps:

1. Isolate the radical expression involving the variable. If more than one radical expression involves the variable, then isolate one of them.
2. Raise both sides of the equation to the index of the radical.
3. If there is still a radical equation, repeat steps 1 and 2; otherwise, solve the resulting equation and check the answer in the original equation.

By raising both sides of an equation to a power, some solutions may have been introduced that do not make the original equation true. These solutions are called **extraneous solutions.** For that reason, it becomes necessary to check all solutions.

Example 1: Solve for x: $\sqrt{3x^2+10x}-5=0$.

Isolate the radical expression.

$$\sqrt{3x^2+10x}=5$$

Raise both sides to the index of the radical; in this case, square both sides.

$$\left(\sqrt{3x^2+10x}\right)^2=(5)^2$$

$$3x^2+10x=25$$

$$3x^2+10x-25=0$$

This quadratic equation now can be solved either by factoring or by applying the quadratic formula.

Applying the quadratic formula,

$$3x^2+10x-25=0$$

$$\begin{aligned}
x &= \frac{-10\pm\sqrt{(10)^2-4(3)(-25)}}{2(3)} \\
&= \frac{-10\pm\sqrt{100+300}}{6} \\
&= \frac{-10\pm\sqrt{400}}{6} \\
&= \frac{-10\pm 20}{6} \\
x &= \frac{5}{3} \text{ or } x=-5
\end{aligned}$$

Now, check the results.

If $x=\frac{5}{3}$,

$$\begin{aligned}
\sqrt{3x^2+10x}-5&=0 \\
\sqrt{3\left(\frac{5}{3}\right)^2+10\left(\frac{5}{3}\right)}-5&\stackrel{?}{=}0 \\
\sqrt{3\left(\frac{25}{9}\right)+\frac{50}{3}}-5&\stackrel{?}{=}0 \\
\sqrt{\frac{25}{3}+\frac{50}{3}}-5&\stackrel{?}{=}0 \\
\sqrt{\frac{75}{3}}-5&\stackrel{?}{=}0 \\
\sqrt{25}-5&\stackrel{?}{=}0 \\
5-5&\stackrel{?}{=}0 \\
0&=0\checkmark
\end{aligned}$$

If $x = -5$,

$$\sqrt{3x^2+10x}-5=0$$

$$\sqrt{3(-5)^2+10(-5)}-5\stackrel{?}{=}0$$

$$\sqrt{3(25)-50}-5\stackrel{?}{=}0$$

$$\sqrt{75-50}-5\stackrel{?}{=}0$$

$$\sqrt{25}-5\stackrel{?}{=}0$$

$$5-5\stackrel{?}{=}0$$

$$0=0\checkmark$$

The solution is $x=\frac{5}{3}$ or $x = -5$.

Example 2: Solve for *x*: $7+\sqrt{a-3}=1$.

Isolate the radical expression.

$$\sqrt{a-3}=-6$$

There is no solution, since $\sqrt{a-3}$ cannot have a negative value.

Example 3: Solve for *x*: $\sqrt{3x-5}+\sqrt{x-1}=2$.

Isolate one of the radical expressions.

$$\sqrt{3x-5}=2-\sqrt{x-1}$$

Raise both sides to the index of the radical; in this case, square both sides.

$$\left(\sqrt{3x-5}\right)^2=\left(2-\sqrt{x-1}\right)^2$$

$$3x-5=\left(2-\sqrt{x-1}\right)\left(2-\sqrt{x-1}\right)$$

$$=4-2\sqrt{x-1}-2\sqrt{x-1}+x-1$$

$$=3-4\sqrt{x-1}+x$$

This is still a radical equation. Isolate the radical expression.

$$3x-5=3-4\sqrt{x-1}+x$$

$$4\sqrt{x-1}=8-2x$$

Raise both sides to the index of the radical; in this case, square both sides.

$$\left(4\sqrt{x-1}\right)^2 = (8-2x)^2$$
$$16(x-1) = 64 - 32x + 4x^2$$
$$16x - 16 = 64 - 32x + 4x^2$$
$$0 = 4x^2 - 48x + 80$$

This can be solved by factoring and by using the quadratic formula. Factor out the 4 first so you can work with smaller numbers. (***Note:*** The quadratic formula would have produced the same results from the original equation.)

$$4(x^2 - 12x + 20) = 0$$

Divide both sides by 4 to get

$$x^2 - 12x + 20 = 0$$

Now use the quadratic formula.

$$x^2 - 12x + 20 = 0$$
$$= \frac{-(-12) \pm \sqrt{(-12)^2 - (4)(1)(20)}}{2(1)}$$
$$= \frac{12 \pm \sqrt{144 - 80}}{2}$$
$$= \frac{12 \pm \sqrt{64}}{2}$$
$$= \frac{12 \pm 8}{2}$$
$$x = 10 \quad \text{or} \quad x = 2$$

Check the solutions.

If $x = 10$,

$$\sqrt{3x-5} + \sqrt{x-1} = 2$$
$$\sqrt{3(10)-5} + \sqrt{10-1} \overset{?}{=} 2$$
$$\sqrt{25} + \sqrt{9} \overset{?}{=} 2$$
$$5 + 3 \overset{?}{=} 2$$
$$8 \neq 2$$

So $x = 10$ is not a solution.

If $x = 2$,

$$\sqrt{3x-5}+\sqrt{x-1}=2$$
$$\sqrt{3(2)-5}+\sqrt{2-1}\stackrel{?}{=}2$$
$$\sqrt{1}+\sqrt{1}\stackrel{?}{=}2$$
$$1+1\stackrel{?}{=}2$$
$$2=2\ \checkmark$$

The only solution is $x = 2$.

Example 4: Solve for x: $\sqrt[3]{2x+3}+5=2$.

Isolate the radical involving the variable.

$$\sqrt[3]{2x+3}=-3$$

Since radicals with odd indexes *can* have negative answers, this problem does have solutions. Raise both sides of the equation to the index of the radical; in this case, cube both sides.

$$\left(\sqrt[3]{2x+3}\right)^3=(-3)^3$$
$$2x+3=-27$$
$$2x=-30$$
$$x=-15$$

Solving Quadratic Inequalities

To solve a quadratic inequality, follow these steps:

1. Solve the inequality as though it were an equation with zero on one side.

   The real solutions to the equation become boundary points on the number line for the solution to the inequality.

2. The boundary points will be solid circles if the original inequality includes equality; otherwise, the boundary points will be open circles.

3. Select "test points" in each of the regions created by the boundary points. Replace these "test points" in the original inequality.
4. If a test point satisfies the original inequality, then the region that contains that test point is part of the solution.
5. Represent the solution in graphic form, in solution set form, and in interval notation.

Example 1: Solve for x: $(x - 3)(x + 2) > 0$.

Solve $(x - 3)(x + 2) = 0$. By the zero-product rule,

$$x-3=0 \quad \text{or} \quad x+2=0$$
$$x=3 \qquad\qquad x=-2$$

Locate the boundary points. Here, the boundary points are open circles because the original inequality does not include equality.

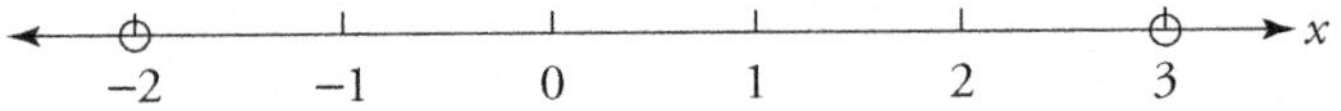

Select "test points" in the different regions created, as shown below.

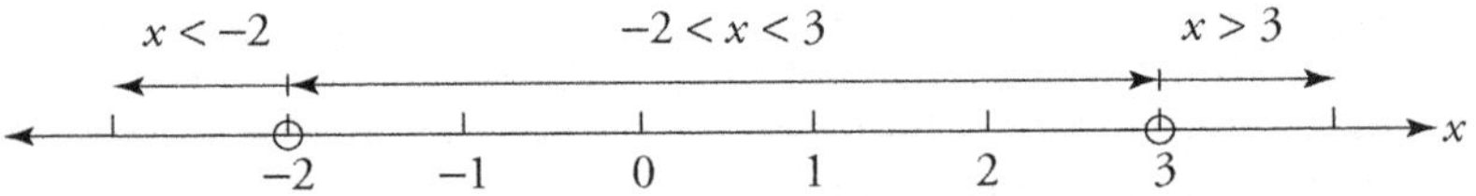

Try $x = -3$, $x = 0$, and $x = 4$. Determine if the test points satisfy the original inequality.

$$x=-3 \qquad\qquad x=0 \qquad\qquad x=4$$

$$(x-3)(x+2)>0 \qquad (x-3)(x+2)>0 \qquad (x-3)(x+2)>0$$

$$(-3-3)(-3+2)\overset{?}{>}0 \qquad (0-3)(0+2)\overset{?}{>}0 \qquad (4-3)(4+2)\overset{?}{>}0$$

$$6>0 \checkmark \qquad\qquad -6\not>0 \qquad\qquad 6>0 \checkmark$$

Since $x = -3$ satisfies the original inequality, the region $x < -2$ is part of the solution. Since $x = 0$ does not satisfy the original inequality, the region $-2 < x < 3$ is not part of the solution. Since $x = 4$ satisfies the original inequality, the region $x > 3$ is part of the solution.

Represent the solution in graphic form, in solution set form, and in interval notation. The graphic form is shown in the following figure.

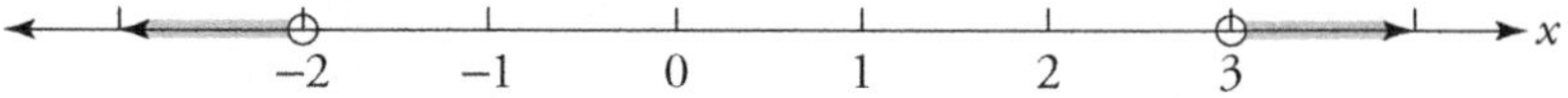

The solution set form is $\{x|x < -2 \text{ or } x > 3\}$.

Interval notation is $(-\infty, -2) \cup (3, \infty)$.

Reminder: When using interval notation, round parentheses indicate exclusion and square brackets indicate inclusion.

Example 2: Solve for x: $9x^2 - 2 \leq -3x$.

$$9x^2 - 2 = -3x$$

$$9x^2 + 3x - 2 = 0$$

By factoring,

$$(3x - 1)(3x + 2) = 0$$

$$3x - 1 = 0 \quad \text{or} \quad 3x + 2 = 0$$

$$x = \frac{1}{3} \qquad\qquad x = -\frac{2}{3}$$

Locate the boundary points using solid dots, as shown in the following figure, since the original inequality includes equality.

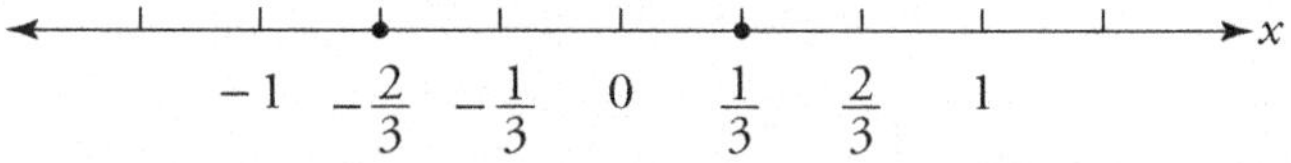

Select "test points" in the regions created.

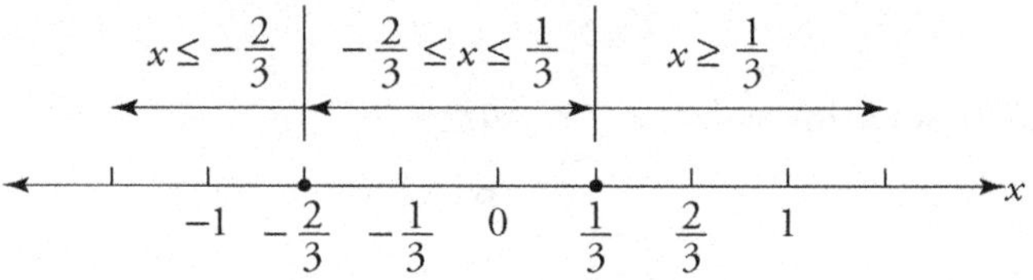

Try $x = -1$, $x = 0$, and $x = 1$. Determine if the test points satisfy the original inequality.

$x=-1$	$x=0$	$x=1$
$9x^2-2\le -3x$	$9x^2-2\le -3x$	$9x^2-2\le -3x$
$9(-1)^2-2\overset{?}{\le}-3(-1)$	$9(0)^2-2\overset{?}{\le}-3(0)$	$9(1)^2-2\overset{?}{\le}-3(1)$
$9-2\overset{?}{\le}3$	$0-2\overset{?}{\le}0$	$9-2\overset{?}{\le}-3$
$7\not\le 3$	$-2\le 0$ ✓	$7\not\le -3$

Since $x = -1$ does not satisfy the original inequality, the region $x\le -\frac{2}{3}$ is not part of the solution. Since $x = 0$ does satisfy the original inequality, the region $-\frac{2}{3}\le x\le \frac{1}{3}$ is part of the solution. Since $x = 1$ does not satisfy the original inequality, the region $x\ge \frac{1}{3}$ is *not* part of the solution.

Represent the solution in graphic form, in solution set form, and in interval notation.

Graphic form:

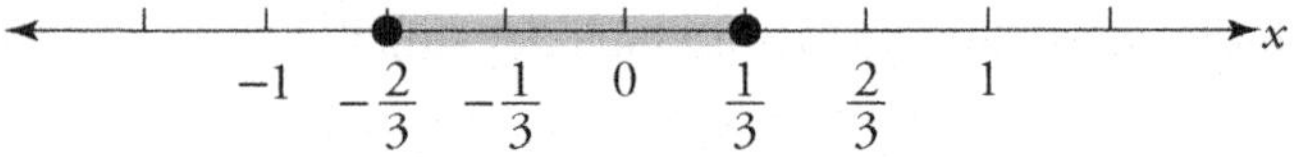

The solution set form is $\left\{x \middle| -\frac{2}{3}\le x\le \frac{1}{3}\right\}$.

Interval notation is $\left[-\frac{2}{3}, \frac{1}{3}\right]$.

Example 3: Solve for *x*: $4t^2 - 9 < -4t$.

$$4t^2-9=-4t$$

$$4t^2+4t-9=0$$

Since this quadratic is not easily factorable, the quadratic formula is used to solve it.

$$t = \frac{-4 \pm \sqrt{(4)^2 - 4(4)(-9)}}{2(4)}$$

$$= \frac{-4 \pm \sqrt{16+144}}{8}$$

$$= \frac{-4 \pm \sqrt{160}}{8}$$

$$= \frac{-4 \pm 4\sqrt{10}}{8}$$

Reduce by dividing out the common factor of 4.

$$t = \frac{-1 \pm \sqrt{10}}{2}$$

$$t = \frac{-1+\sqrt{10}}{2} \quad \text{or} \quad t = \frac{-1-\sqrt{10}}{2}$$

Since $\sqrt{10}$ is approximately 3.2,

$$t \approx \frac{-1+3.2}{2} \quad \text{or} \quad t \approx \frac{-1-3.2}{2}$$

$$\approx 1.1 \qquad\qquad \approx -2.1$$

Locate the boundary points using open circles, as shown in the following figure, since the original inequality does not include equality.

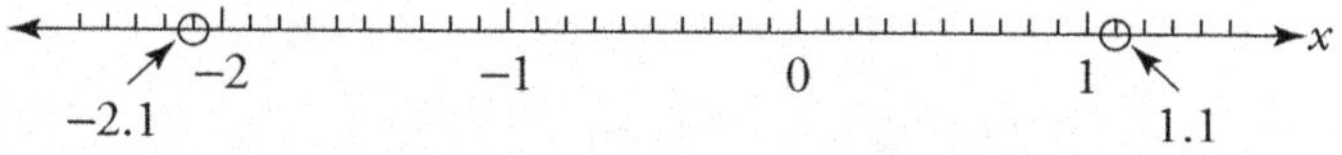

Select "test points" in the different regions created.

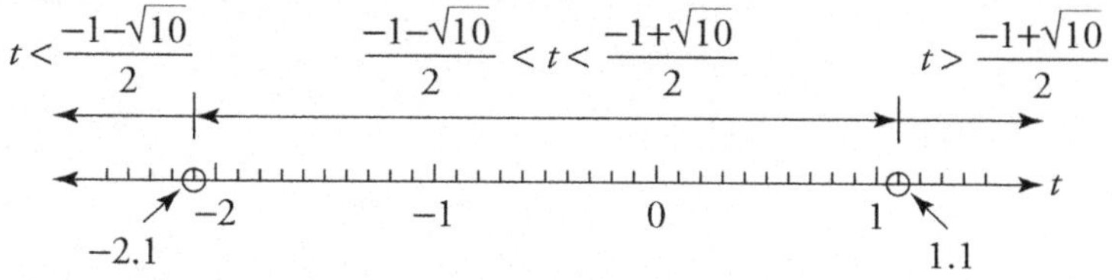

Try $t = -3$, $t = 0$, and $t = 2$. Determine if the test points satisfy the original inequality.

$$t=-3 \qquad t=0 \qquad t=2$$

$$4t^2-9<-4t \qquad 4t^2-9<-4t \qquad 4t^2-9<-4t$$

$$4(-3)^2-9\overset{?}{<}-4(-3) \qquad 4(0)^2-9\overset{?}{<}-4(0) \qquad 4(2)^2-9\overset{?}{<}-4(2)$$

$$27\not<12 \qquad -9<0\ \checkmark \qquad 7\not<-8$$

Since $t = -3$ does not satisfy the original inequality, the region $t<\frac{-1-\sqrt{10}}{2}$ is not part of the solution. Since $t = 0$ does satisfy the original inequality, the region $\frac{-1-\sqrt{10}}{2}<t<\frac{-1+\sqrt{10}}{2}$ is part of the solution. Since $t = 2$ does not satisfy the original inequality, the region $t>\frac{-1+\sqrt{10}}{2}$ is not part of the solution.

Represent the solution in graphic form, in a solution set form, and in interval notation.

Graphic form:

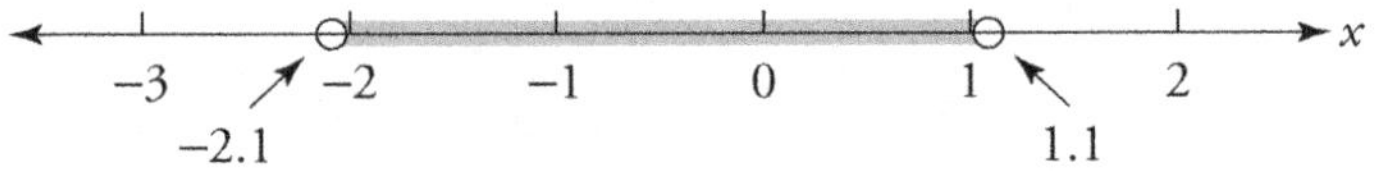

The solution set form is $\left\{t \middle| \frac{-1-\sqrt{10}}{2}<t<\frac{-1+\sqrt{10}}{2}\right\}$.

Interval notation is $\left(\frac{-1-\sqrt{10}}{2}, \frac{-1+\sqrt{10}}{2}\right)$.

Example 4: Solve for x: $x^2 + 2x + 5 < 0$.

$$x^2 + 2x + 5 = 0$$

Since this quadratic is not factorable using rational numbers, the quadratic formula will be used to solve it.

$$\begin{aligned} x &= \frac{-2 \pm \sqrt{(2)^2 - 4(1)(5)}}{2(1)} \\ &= \frac{-2 \pm \sqrt{4-20}}{2} \\ &= \frac{-2 \pm \sqrt{-16}}{2} \\ &= \frac{-2 \pm 4i}{2} \\ &= -1 \pm 2i \end{aligned}$$

Substitute $x = -1 \pm 2i$ into $x^2 + 2x + 5 < 0$.

$$(-1 \pm 2i)^2 + 2(-1 \pm 2i) + 5 < 0$$

$$1 \pm \cancel{2i} \pm \cancel{2i} + 4i^2 - 2 \pm \cancel{4i} + 5 \overset{?}{<} 0$$

$$1 + 4i^2 - 2 + 5 \overset{?}{<} 0$$

$$1 - 4 - 2 + 5 \overset{?}{<} 0$$

$$-3 - 2 + 5 \overset{?}{<} 0$$

$$0 \not< 0$$

The quadratic equation has no real solutions. Also, the quadratic inequality has no real solutions since $0 < 0$ is not a true statement.

These are imaginary answers and cannot be graphed on a real number line. Therefore, the inequality $x^2 + 2x + 5 < 0$ has no real solutions.

Example 5: Solve for x: $x^2 + 2x + 5 > 0$.

Since any value of x (positive, negative, or zero) will keep the expression positive, the solution is *all real numbers*. Therefore, this will make the inequality true.

Chapter Check-Out

Questions

1. Solve for x by factoring: $3x^2 - 10x + 8 = 0$.
2. Solve for x by completing the square: $x^2 - 6x - 11 = 0$.
3. Solve for x by using the quadratic formula: $2x^2 - 3x - 8 = 0$.
4. Find the value of the discriminant in each of the following equations and explain the nature of the roots:
   (a) $x^2 + 6x + 9 = 0$
   (b) $x^2 + 7x + 12 = 0$
   (c) $x^2 + 5x + 10 = 0$
5. A number is three more than twice a second number. The sum of their squares is 137. Find the numbers.
6. A frame of uniform width surrounds a picture that is 16 by 12 inches. The frame has an area of 165 square inches. Find the width of the frame.

Answers

1. $x = 2$ or $x = \dfrac{4}{3}$
2. $x = 3 \pm 2\sqrt{5}$
3. $x = \dfrac{3 \pm \sqrt{73}}{4}$
4. **(a)** 0; one real solution; **(b)** 1; two real solutions;
   (c) –15; two imaginary solutions
5. The equation is $x^2 + (2x + 3)^2 = 137$, so $x = 4$. The two positive numbers are, therefore, 4 and 11.
6. The equation for the area if the frame is $(16 + 2x)(12 + 2x) - 16(12) = 165$. Solving the equation for x, the width of the frame, $x = 2.5$ inches.

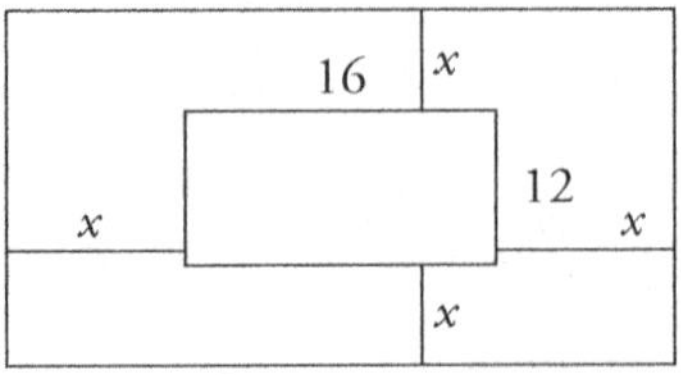

Chapter 12
CONIC SECTIONS

Chapter Check-In

- ❑ Defining the four conic sections
- ❑ Analyzing equations of circles
- ❑ Determining properties of parabolas
- ❑ Analyzing equations of ellipses
- ❑ Determining asymptotes of hyperbolas

Common Core Standard: Expressing Geometric Properties with Equations

Identify, graph, and write equations of all four conic sections. Translate between the geometric description and the equation for a conic section (G.GPE).

Conic sections are formed in a plane when that plane slices through one or both of a pair of right circular cones stacked tip to tip. Whether the result is a circle, ellipse, parabola, or hyperbola depends only upon the angle at which the plane slices through the cone(s). Conic sections are described mathematically by quadratic equations—some of which contain more than one variable. The concept of symmetry is applied to each of the four conic sections.

The Four Conic Sections

When a single or stacked pair of right circular cones is sliced by a plane, the curved cross section formed by the plane and cone is called a **conic section.** The four main conic sections are the circle, the parabola, the ellipse, and the hyperbola, as shown in the following figure.

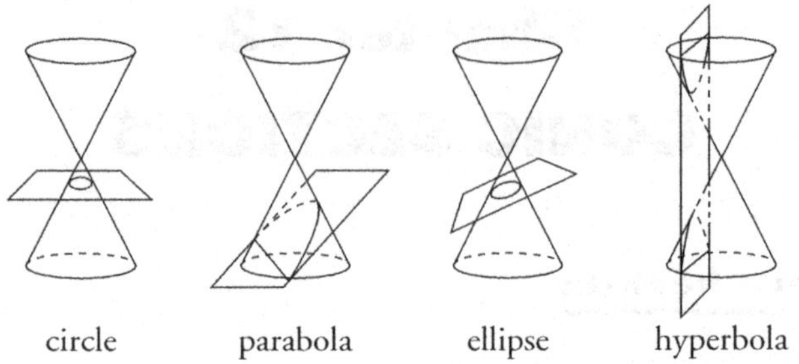

Circles

A **circle** is the set of points in a plane that are equidistant from a given point. That point is called the **center** of the circle, and the distance from it to any point on the circle is called the **radius** of the circle. The standard form for the equation of a circle with its center at (0, 0) and with a radius of length r is represented by the following equation:

$$x^2 + y^2 = r^2$$

Example 1: Graph $x^2 + y^2 = 16$.

Recognize that $x^2 + y^2 = 16$ is the equation of a circle centered at (0, 0) with $r = 4$, as shown in the following figure.

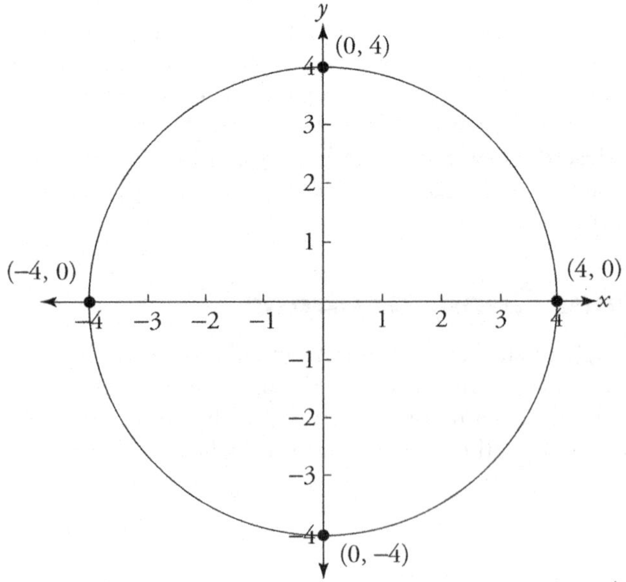

Example 2: Find the standard form for the equation of the circle centered at (0, 0) with a radius of $\sqrt{5}$.

The standard form for a circle centered at (0, 0) with a radius of r is $x^2 + y^2 = r^2$.

Replacing r with $\sqrt{5}$, the equation becomes

$$x^2 + y^2 = \left(\sqrt{5}\right)^2$$

$$x^2 + y^2 = 5$$

Therefore, $x^2 + y^2 = 5$ is the standard form of the equation of a circle centered at (0, 0) with a radius of $\sqrt{5}$.

The standard form of the equation of a circle centered at (h, k) with a radius of r is $(x - h)^2 + (y - k)^2 = r^2$.

Note that when $(h, k) = (0, 0)$, that leads to

$$(x-0)^2 + (y-0)^2 = r^2$$

$$x^2 + y^2 = r^2$$

This is the equation for a circle centered at the origin.

Example 3: Graph the equation $(x - 3)^2 + (y + 2)^2 = 25$.

This equation represents a circle centered at (3, −2) with a radius of 5, as shown in the following figure.

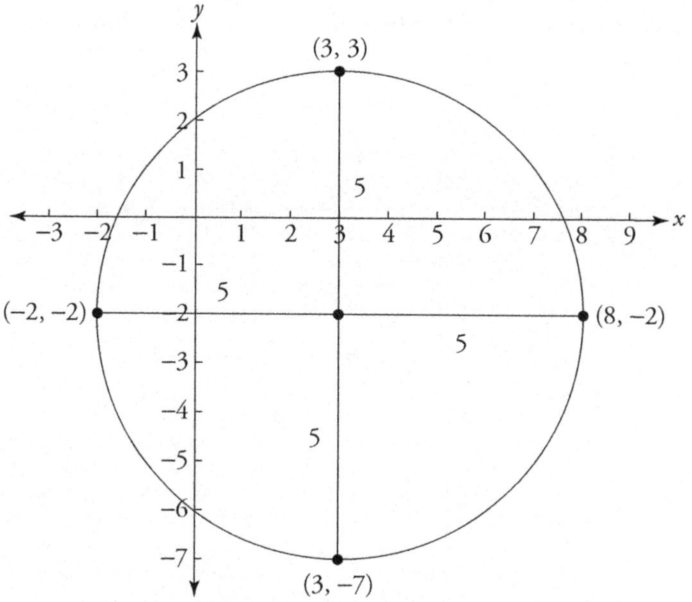

Example 4: Find the standard form for the equation of the circle centered at $(-6, 2)$ with a radius of $3\sqrt{2}$.

The standard form for the equation of a circle centered at (h, k) with radius r is $(x - h)^2 + (y - k)^2 = r^2$.

Replacing h with -6, k with 2, and r with $3\sqrt{2}$, the equation becomes

$$[x-(-6)]^2+(y-2)^2=\left(3\sqrt{2}\right)^2$$

Therefore, $(x + 6)^2 + (y - 2)^2 = 18$, which is the standard form of the equation of the circle centered at $(-6, 2)$ with radius of $3\sqrt{2}$.

Example 5: From the equation $x^2 + y^2 - 8x + 12y - 12 = 0$, find the center and radius for the circle. Then graph the circle.

This equation can be rewritten as

$$x^2 - 8x + y^2 + 12y = 12$$

Now, complete the square for each variable and add those amounts to both sides of the equation.

$$x^2-8x+\underline{16}+y^2+12y+\underline{36}=12+\underline{16}+\underline{36}$$
$$(x-4)^2+(y+6)^2=64$$

This circle is centered at $(4, -6)$ with a radius of 8, as shown in the following figure.

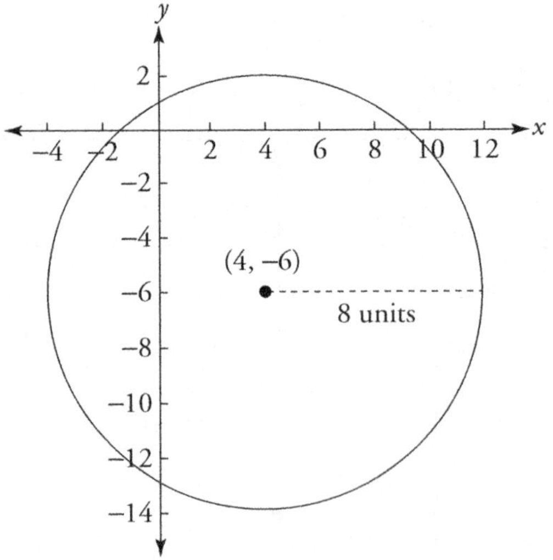

Parabolas

A **parabola** is the set of points in a plane that are the same distance (equidistant) from a given point and a given line in that plane. The given point is called the **focus,** and the given line is called the **directrix.** The midpoint of the perpendicular segment from the focus to the directrix is called the **vertex** of the parabola. The line that passes through the vertex and the focus is called the **axis of symmetry of the parabola**. The figures below illustrate the two possible parabolas.

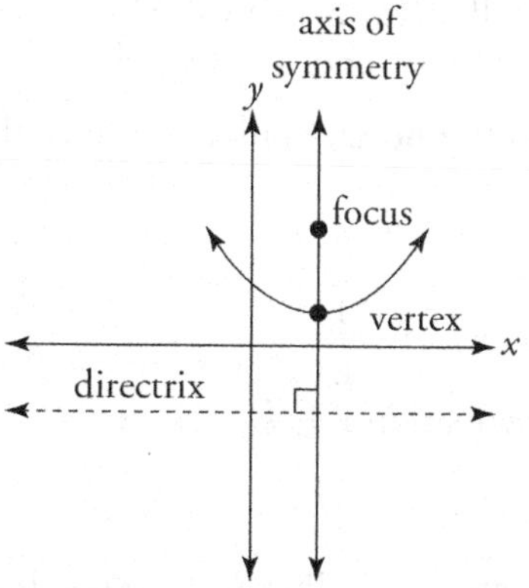

(a)

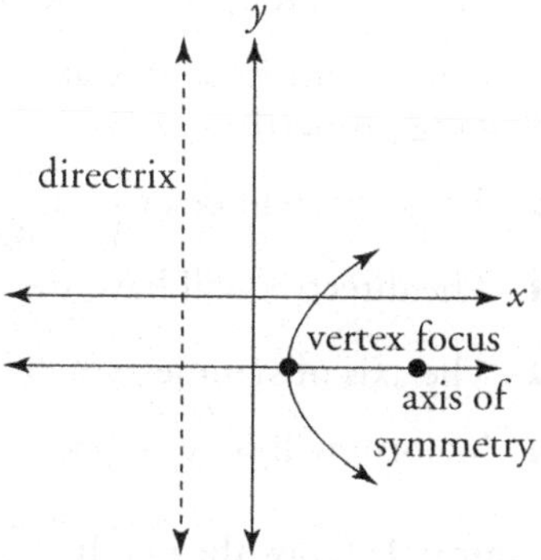

(b)

The equations of these parabolas can be written in vertex form:

- **Form 1:** $y = a(x - h)^2 + k$, Figure (a) above
- **Form 2:** $x = a(y - k)^2 + h$, Figure (b) above

Note: The equation of a parabola is squared only in one variable—the variable squared has a directrix that is parallel to that axis. In Form 1, the parabola opens vertically. (It opens in the "y" direction.) If $a > 0$, it opens upward. If $a < 0$, it opens downward. The distances from the vertex to the focus and from the vertex to the directrix line are the same. This distance is defined as

$$\left|\frac{1}{4a}\right|$$

A parabola with its vertex at (h, k), opening vertically, will have the following properties:

- The focus will be at $\left(h,\ k+\frac{1}{4a}\right)$.
- The directrix will have the equation $y = k - \frac{1}{4a}$.
- The axis of symmetry will have the equation $x = h$.
- Its form will be $y = a(x - h)^2 + k$.

In Form 2, the parabola opens horizontally. (It opens in the "x" direction.) If $a > 0$, it opens to the right. If $a < 0$, it opens to the left.

A parabola with its vertex at (h, k), opening horizontally, will have the following properties.

- The focus will be at $\left(h+\frac{1}{4a},\ k\right)$.
- The directrix will have the equation $x = h - \frac{1}{4a}$.
- The axis of symmetry will have the equation $y = k$.
- Its form will be $x = a(y - k)^2 + h$.

Example 1: Draw the graph of $y = x^2$. State which direction the parabola opens and determine its vertex, focus, directrix, and axis of symmetry.

The equation $y = x^2$ can be written as $y = 1(x - 0)^2 + 0$; $a = 1$, $h = 0$, and $k = 0$. Since $a > 0$ and the parabola opens vertically, its direction is up.

Vertex: $(h, k) = (0, 0)$

$$\text{Focus: } \left(h,\ k+\frac{1}{4a}\right) = \left(0,\ 0+\frac{1}{4(1)}\right)$$
$$= \left(0,\ \frac{1}{4}\right)$$

$$\text{Directrix: } y = k - \frac{1}{4a}$$
$$y = 0 - \frac{1}{4}$$
$$y = -\frac{1}{4}$$

$$\text{Axis of symmetry: } x = h$$
$$x = 0$$

The figure below illustrates properties of the parabola $y = x^2$.

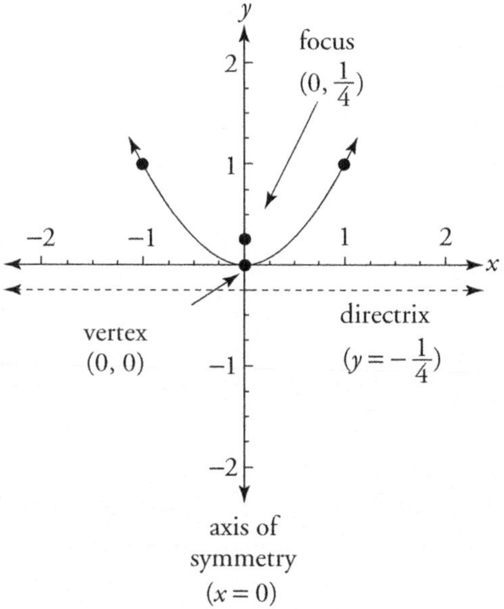

Example 2: Graph $x = -\frac{1}{8}(y+2)^2 - 3$. State which direction the parabola opens and determine its vertex, focus, directrix, and axis of symmetry.

The equation $x = -\frac{1}{8}(y+2)^2 - 3$ is the same as $x = -\frac{1}{8}[(y-(-2)]^2 + (-3)$; $a = -\frac{1}{8}$, $h = -3$, and $k = -2$.

Since $a < 0$ and the parabola opens horizontally, this parabola opens to the left.

Vertex: $(h, k) = (-3, -2)$

$$\text{Focus: } \left(h + \frac{1}{4a}, k\right) = \left(-3 + \frac{1}{4\left(-\frac{1}{8}\right)}, -2\right)$$

$$= [-3 + (-2), -2]$$

$$= (-5, -2)$$

Directrix: $x = h - \frac{1}{4a}$

$$x = -3 - \frac{1}{4\left(-\frac{1}{8}\right)}$$

$$x = -3 - (-2)$$

$$x = -1$$

Axis of symmetry: $y = k$

$$y = -2$$

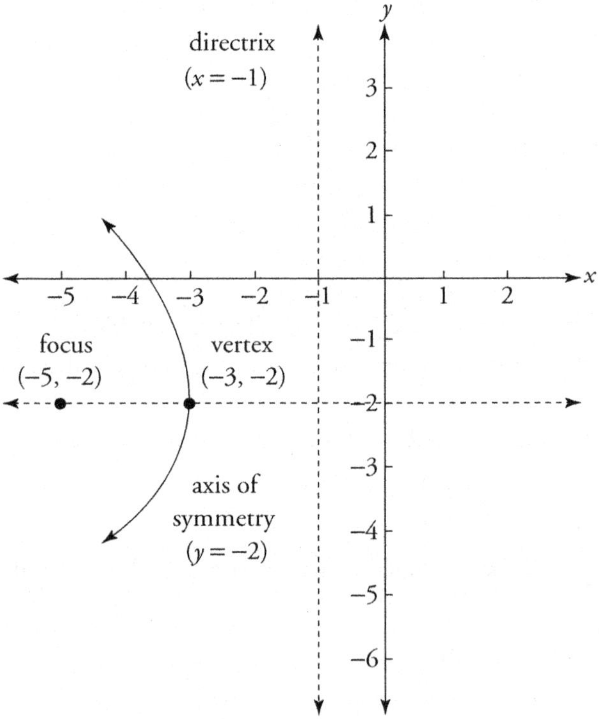

Example 3: Write the equation $x = 5y^2 - 30y + 11$ in vertex form: $x = a(y - k)^2 + h$. Determine the direction of opening, vertex, focus, directrix, and axis of symmetry.

Factor out the coefficient of y^2 from the terms involving y to complete the square.

$$x = 5(y^2 - 6y) + 11$$

Completing the square within the parentheses adds $5(9) = 45$ to both sides of the equation.

$$x+45=5(y^2-6y+9)+11$$
$$x+45=5(y-3)^2+11$$

Subtract 45 from both sides to have vertex form.

$$x = 5(y - 3)^2 - 34$$

Direction: Opens to the right ($a > 0$, opens horizontally)

Vertex: $(h, k) = (-34, 3)$

Focus: $\left(h+\frac{1}{4a}, k\right)=\left(-34+\frac{1}{20}, 3\right)$

$$=\left(-33\frac{19}{20}, 3\right)$$

Directrix: $x = h - \frac{1}{4a}$

$$x=-34-\frac{1}{20}$$

$$x=-34\frac{1}{20}$$

Axis of symmetry: $y = k$

$$y=3$$

Tip: This process of completing the square in one of the variables allows finding the vertex of any parabola. The vertex is referred to as a *minimum* or a *maximum point* when the parabola opens vertically. This becomes a valuable tool in finding minimum and maximum values in word problems.

Example 4: Peyton has 72 feet of fencing to build a dog pen along one wall of his house. Find the maximum width that the dog pen can be, and find the maximum area that can be fenced in.

The pen is a rectangle whose area is $A = lw$. If w is the width, then the length would be $72 - 2w$.

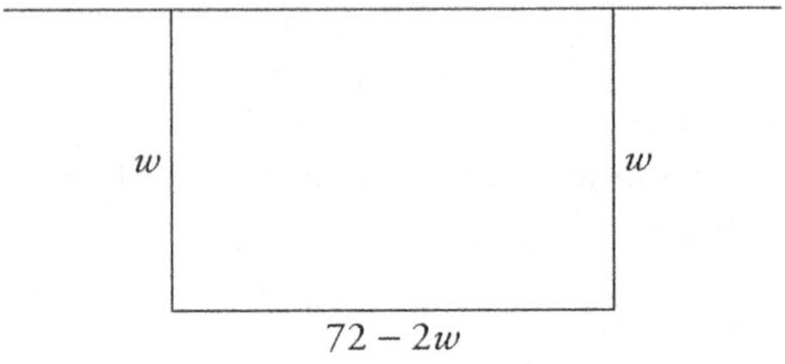

Substituting:

$$A = (72 - 2w)w$$
$$A = 72w - 2w^2$$
$$A = -2(-36w + w^2)$$
or
$$A = -2(w^2 - 36w)$$

Complete the square:

$$A = -2\left(w^2 - 36w + 324\right)$$

Subtract 648 from each side $(-2)(-324) = -648$:

$$A - 648 = -2(w - 18)^2$$
$$A = -2(w - 18)^2 + 648$$

After completing the square, this quadratic is now in vertex form. This equation represents the graph of a vertical parabola that opens down and has a vertex of (18, 648).

The interpretation of this point is a *maximum*. The maximum width value is 18 feet, and the maximum area is 648 square feet.

Ellipses

An **ellipse** is the set of points in a plane such that the sum of the distances from two fixed points in that plane stays constant. The two points are each called a **focus.** The plural of focus is **foci.** The midpoint of the segment joining the foci is called the **center** of the ellipse. An ellipse has two axes of symmetry. The longer one is called the **major axis,** and the shorter one is called the **minor axis.** The two axes intersect perpendicularly at the center of the ellipse.

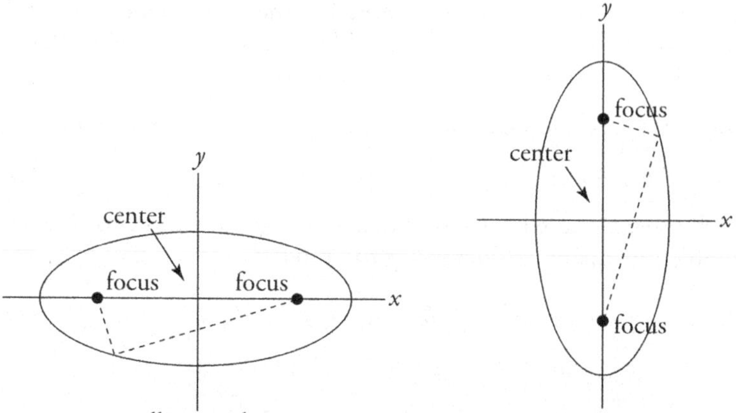

ellipse with a horizontal major axis

(a)

ellipse with a vertical major axis

(b)

Major axis along the *x*-axis or along the *y*-axis

The equation of an ellipse that is centered at (0, 0) and has its major axis along the *x*-axis has the following standard form:

$$\frac{x^2}{a^2}+\frac{y^2}{b^2}=1 \quad \left(a^2>b^2\right)$$

The length of the major axis is $2|a|$ and the length of the minor axis is $2|b|$. The endpoints of the major axis are $(a, 0)$ and $(-a, 0)$ and are referred to as the **major intercepts.** The endpoints of the minor axis are $(0, b)$ and $(0, -b)$ and are referred to as the **minor intercepts.** If $(c, 0)$ and $(-c, 0)$ are the locations of the foci, then c can be found using the following equation:

$$c^2 = a^2 - b^2$$

If an ellipse has its major axis along the *y*-axis and is centered at (0, 0), the standard form becomes

$$\frac{x^2}{b^2}+\frac{y^2}{a^2}=1 \quad \left(a^2>b^2\right)$$

The endpoints of the major axis become $(0, a)$ and $(0, -a)$. The endpoints of the minor axis become $(b, 0)$ and $(-b, 0)$. The foci always lie on the major axis. They are located at $(0, c)$ and $(0, -c)$, with $c^2 = a^2 - b^2$.

When an ellipse is written in standard form, the major axis direction is determined by noting which variable has the larger denominator. The major axis either lies along that variable's axis or is parallel to that variable's axis, depending on where the center is located.

Example 1: Graph the following ellipse: $\frac{x^2}{4}+\frac{y^2}{9}=1$. Find its major intercepts, length of the major axis, minor intercepts, length of the minor axis, and foci.

This ellipse is centered at (0, 0). Since the larger denominator is with the y variable, the major axis lies along the y-axis.

$$a^2 = 9 \qquad b^2 = 4 \qquad c^2 = a^2 - b^2$$
$$|a| = 3 \qquad |b| = 2 \qquad = 9 - 4$$
$$|c| = \sqrt{5}$$

Major intercepts: (0, 3), (0, −3)
Length of major axis: $2|a| = 6$
Minor intercepts: (2, 0), (−2, 0)
Length of minor axis: $2|b| = 4$
Foci: $(0, \sqrt{5})$, $(0, -\sqrt{5})$

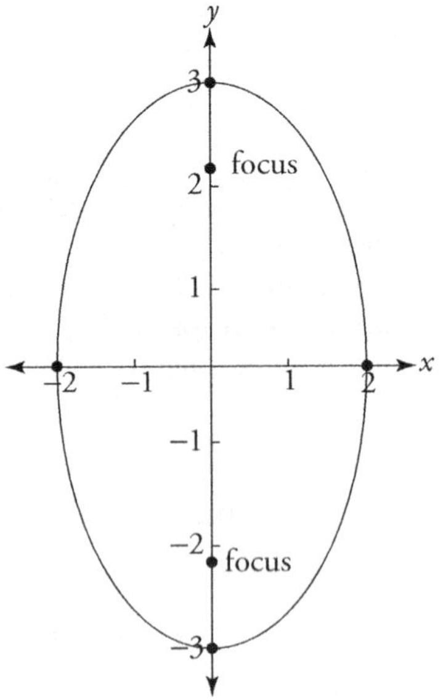

Example 2: Graph the following ellipse: $4x^2 + 25y^2 = 100$. Find its major and minor intercepts and its foci.

Write $4x^2 + 25y^2 = 100$ in standard form by dividing each side by 100.

$$\frac{4x^2}{100} + \frac{25y^2}{100} = \frac{100}{100}$$
$$\frac{x^2}{25} + \frac{y^2}{4} = 1$$

This ellipse is centered at (0, 0). Since the larger denominator is with the x variable, the major axis lies along the x-axis.

$$a^2 = 25 \qquad b^2 = 4 \qquad c^2 = a^2 - b^2$$
$$|a| = 5 \qquad |b| = 2 \qquad = 25 - 4$$
$$= 21$$
$$|c| = \sqrt{21}$$

Major intercepts: (5, 0), (−5, 0)
Minor intercepts: (0, 2), (0, −2)
Foci: $(\sqrt{21}, 0)$, $(-\sqrt{21}, 0)$

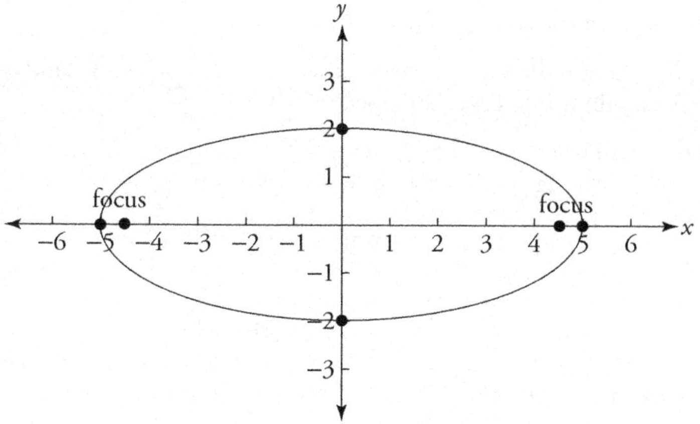

Major axis parallel to the *x*-axis or to the *y*-axis

The standard form for an ellipse centered at (h, k) with its major axis parallel to the x-axis is

$$\frac{(x-h)^2}{a^2}+\frac{(y-k)^2}{b^2}=1 \quad (a^2>b^2)$$

Major intercepts: $(h + a, k)$, $(h - a, k)$
Minor intercepts: $(h, k + b)$, $(h, k - b)$
Foci: $(h + c, k)$, $(h - c, k)$ with $c=\sqrt{a^2-b^2}$

The standard form for an ellipse centered at (h, k) with its major axis parallel to the y-axis is

$$\frac{(x-h)^2}{b^2}+\frac{(y-k)^2}{a^2}=1 \quad (a^2>b^2)$$

Major intercepts: $(h, k + a)$ $(h, k - a)$
Minor intercepts: $(h + b, k)$ $(h - b, k)$
Foci: $(h, k + c)$, $(h, k - c)$ with $c=\sqrt{a^2-b^2}$

The points where the major axis intersects the ellipse are also known as the ellipse's vertices. That means that each major intercept is also known as a **vertex of the ellipse.** Notice that the vertices, foci, and center of an ellipse all have the same x-coordinate when the ellipse's major axis is parallel to the y-axis, whereas the vertices, foci, and center all have the same y-coordinate when the major axis parallels the x-axis. Ellipses may have axes at oblique angles as well as horizontal and vertical, but their study is beyond the scope of Algebra II.

Example 3: Graph the following ellipse: $\frac{(x-2)^2}{36}+\frac{(y+1)^2}{25}=1$. Find its center, vertices, minor intercepts, and foci.

Center: $(2, -1)$

$$a^2=36 \qquad b^2=25 \qquad c^2=a^2-b^2$$
$$|a|=6 \qquad |b|=5 \qquad =36-25$$
$$=11$$
$$|c|=\sqrt{11}$$

Vertices: $(2+6,-1)=(8,-1)$
$(2-6,-1)=(-4,-1)$

Minor intercepts: $(2, -1+5) = (2, 4)$

$$(2, -1-5) = (2, -6)$$

Foci: $\left(2+\sqrt{11}, -1\right), \left(2-\sqrt{11}, -1\right)$

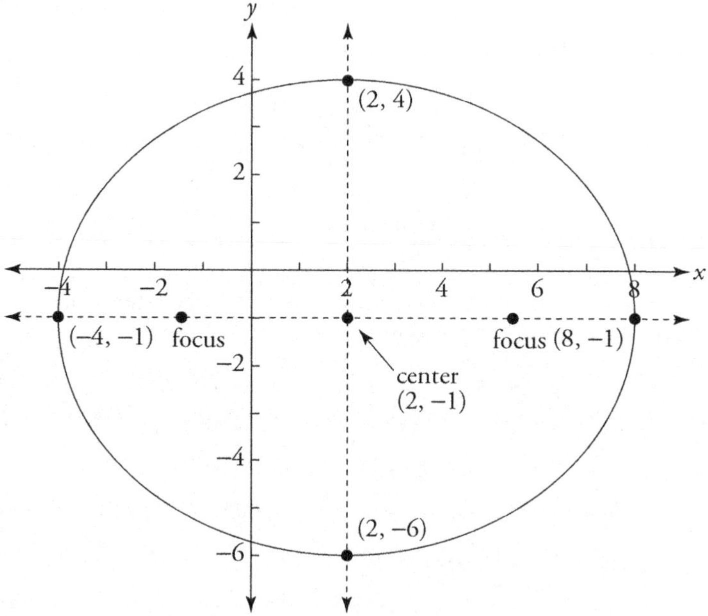

Example 4: Graph the following ellipse: $16x^2 + 25y^2 + 32x - 150y = 159$. Find the coordinates of its center, major and minor intercepts, and foci.

Rearrange terms to get the x terms together and the y terms together for potential factoring.

$$16x^2 + 32x + 25y^2 - 150y = 159$$

Factor out the coefficient of each of the squared terms.

$$16(x^2 + 2x) + 25(y^2 - 6y) = 159$$

Complete the square within each set of parentheses and add the same amount to both sides of the equation.

$$16(x^2+2x+\underline{1})+25(y^2-6y+\underline{9})=159+\underline{16(1)}+\underline{25(9)}$$
$$16(x+1)^2+25(y-3)^2=400$$

Divide each side by 400.

$$\frac{\overset{1}{\cancel{16}}(x+1)^2}{\underset{25}{\cancel{400}}}+\frac{\overset{1}{\cancel{25}}(y-3)^2}{\underset{16}{\cancel{400}}}=\frac{400}{400}$$
$$\frac{(x+1)^2}{25}+\frac{(y-3)^2}{16}=1$$

Center $(-1, 3)$: Since the x variable has the larger denominator, the major axis is parallel to the x-axis.

$$a^2=25 \qquad b^2=16 \qquad c^2=a^2-b^2$$
$$|a|=5 \qquad |b|=4 \qquad =25-16$$
$$=9$$
$$|c|=3$$

Major intercepts: $(-1+5,\ 3)=(4,\ 3)$
$(-1-5,\ 3)=(-6,\ 3)$

Minor intercepts: $(-1,\ 3+4)=(-1,\ 7)$
$(-1,\ 3-4)=(-1,\ -1)$

Foci: $(-1+3,\ 3)=(2,\ 3)$
$(-1-3,\ 3)=(-4,\ 3)$

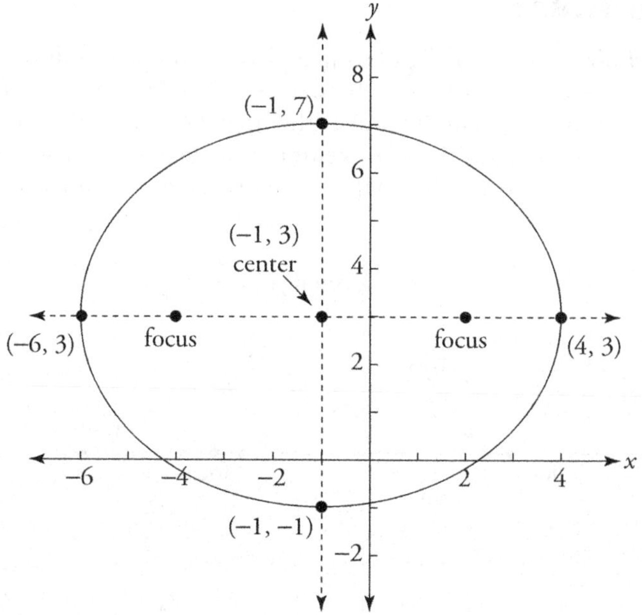

Eccentricity

Ellipses are eccentric, a property that is expressed as a number between 0 and 1. If the eccentricity of an ellipse were 0 (which it cannot be), that ellipse would be a circle. The greater the eccentricity of an ellipse (the closer it is to 1), the more oval in shape the ellipse is. The eccentricity of an ellipse, *e*, may be calculated as the ratio of *c* to *a*, or $e = \frac{c}{a}$.

Recall that $c = \sqrt{a^2 - b^2}$. Consider the last ellipse you graphed:

$$\frac{(x+1)^2}{25} + \frac{(y-3)^2}{16} = 1$$

Now, find its eccentricity:

$$\begin{aligned} e &= \frac{c}{a} \\ &= \frac{3}{5} \\ &= 0.6 \end{aligned}$$

Since $e = 0.6$, and 0.6 is closer to 1 than it is to 0, the ellipse in question is much more oval than it is round.

Hyperbolas

A **hyperbola** is the set of all points in a plane such that the absolute value of the difference of the distances between two fixed points stays constant. The two given points are the **foci** of the hyperbola, and the midpoint of the segment joining the foci is the **center** of the hyperbola. The hyperbola looks like two opposing "U-shaped" curves, as shown in the following figure.

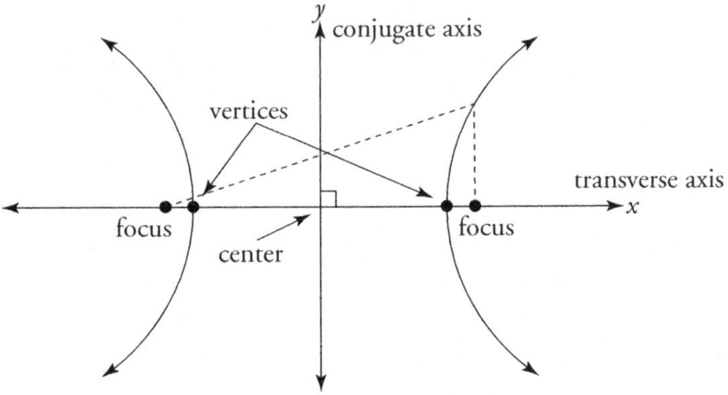

A hyperbola has two axes of symmetry (refer to the previous figure). The axis along the direction the hyperbola opens is called the **transverse axis.** The **conjugate axis** passes through the center of the hyperbola and is perpendicular to the transverse axis. The points of intersection of the hyperbola and the transverse axis are called the *vertices* (plural of **vertex**) of the hyperbola.

A hyperbola centered at (0, 0) whose transverse axis is along the x-axis has the following equation as its standard form:

$$\frac{x^2}{a^2} - \frac{y^2}{b^2} = 1$$

where $(a, 0)$ and $(-a, 0)$ are the vertices and $(c, 0)$ and $(-c, 0)$ are its foci.

In the hyperbola, $c = \sqrt{a^2 + b^2}$. (Do not confuse this with the ellipse formula, $c = \sqrt{a^2 - b^2}$.)

As points on a hyperbola get farther from its center, they get closer and closer to two lines called **asymptote lines.** The asymptote lines are used as guidelines in sketching the graph of a hyperbola. To graph the asymptote lines, form a rectangle by using the points $(-a, b)$, $(-a, -b)$, (a, b), and $(a, -b)$ and draw its diagonals as extended lines.

Centered at (0, 0) with transverse axis along the x-axis

For the hyperbola centered at (0, 0) whose transverse axis is along the x-axis, the equation of the asymptote lines becomes

$$y = \pm\frac{b}{a}x \text{ (recall slope-intercept form of a line)}$$

Example 1: Graph the following hyperbola: $\frac{x^2}{16} - \frac{y^2}{25} = 1$. Find its center, vertices, foci, and the equations of its asymptote lines.

This is a hyperbola with center at (0, 0) and its transverse axis along the x-axis.

$$a^2 = 16 \qquad b^2 = 25 \qquad c = \sqrt{a^2 + b^2}$$

$$|a| = 4 \qquad |b| = 5 \qquad = \sqrt{16 + 25}$$

$$= \sqrt{41}$$

$$|c| = \sqrt{41}$$

Vertices: (−4, 0), (4, 0)
Foci: $\left(-\sqrt{41}, 0\right)$, $\left(\sqrt{41}, 0\right)$
Equations of asymptote lines: $y = \pm\frac{5}{4}x$

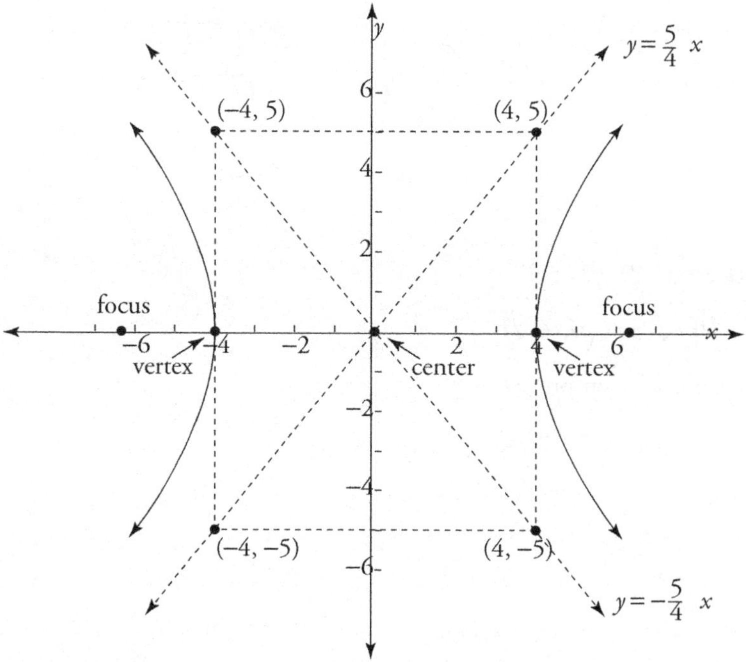

Centered at (0, 0) with transverse axis along the y-axis

A hyperbola centered at (0, 0) whose transverse axis is along the y-axis has the following equation as its standard form:

$$\frac{y^2}{a^2}-\frac{x^2}{b^2}=1$$

The vertices are now $(0, a)$ and $(0, -a)$. The foci are at $(0, c)$ and $(0, -c)$, with $c^2 = a^2 + b^2$. The asymptote lines have equations

$$y=\pm\frac{a}{b}x$$

In general, when a hyperbola is written in standard form, the transverse axis is along, or parallel to, the axis of the variable that appears first.

Example 2: Graph the following hyperbola: $\frac{y^2}{4}-\frac{x^2}{9}=1$. Find its center, vertices, foci, and equations of the asymptote lines.

This is a hyperbola with its center at (0, 0) and its transverse axis along the y-axis, since the y variable appears first.

$$a^2 = 4 \qquad b^2 = 9 \qquad c=\sqrt{a^2+b^2}$$

$$|a| = 2 \qquad |b| = 3 \qquad =\sqrt{4+9}$$

$$=\sqrt{13}$$

$$|c|=\sqrt{13}$$

Vertices: (0, 2), (0, −2)

Foci: $\left(0, \sqrt{13}\right)$, $\left(0, -\sqrt{13}\right)$

Equations of asymptote lines: $y=\pm\frac{2}{3}x$

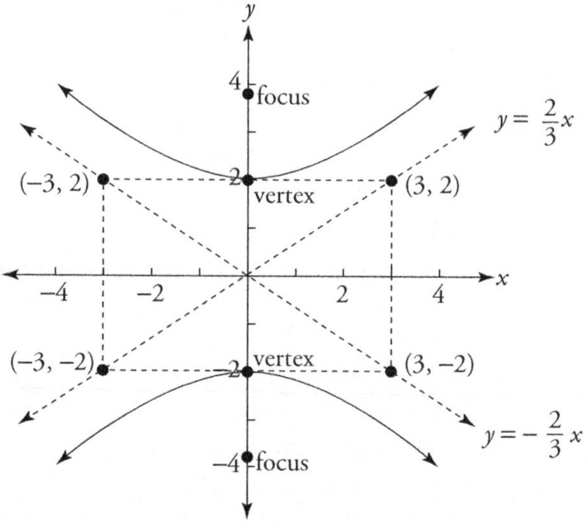

Centered at (*h*, *k*) with a horizontal or vertical transverse axis

A hyperbola centered at (h, k) will have the following standard equations:

- If the transverse axis is horizontal, then

$$\frac{(x-h)^2}{a^2} - \frac{(y-k)^2}{b^2} = 1$$

In this case, the vertices are at $(h + a, k)$ and $(h - a, k)$. The foci are at $(h + c, k)$ and $(h - c, k)$, where $c = \sqrt{a^2 + b^2}$.

- If the transverse axis is vertical, then

$$\frac{(y-k)^2}{a^2} - \frac{(x-h)^2}{b^2} = 1$$

The vertices are at $(h, k + a)$ and $(h, k - a)$, and the foci are at $(h, k + c)$ and $(h, k - c)$, where $c = \sqrt{a^2 + b^2}$.

Example 3: Graph the following hyperbola: $\dfrac{(x-2)^2}{16}-\dfrac{(y+3)^3}{25}=1$.

Center: $(2, -3)$

The transverse axis is horizontal.

Vertices: $(2+4, -3)=(6, -3)$

$(2-4, -3)=(-2, -3)$

Find the coordinates of the corners of the rectangle to help make the asymptote lines.

$$(2+4,\ -3+5)=(6,\ 2)$$
$$(2-4,\ -3+5)=(-2,\ 2)$$
$$(2+4,\ -3-5)=(6,\ -8)$$
$$(2-4,\ -3-5)=(-2,\ -8)$$

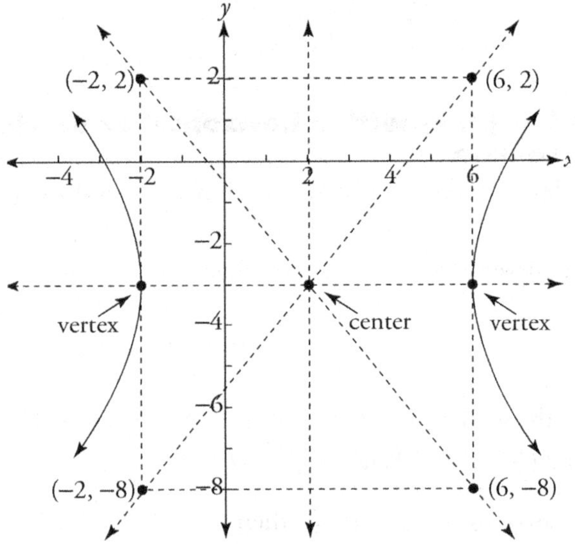

From the equation, the center is (2, –3). The transversal is horizontal because x comes first in the equation.

$a^2=16$	$b^2=25$	$c=\sqrt{a^2+b^2}$
$\lvert a\rvert=4$	$\lvert b\rvert=5$	$c=\sqrt{16+25}$
		$c=\sqrt{41}$

Centered at (0, 0) with a slant transverse axis

The equation $xy = 16$ also represents a hyperbola. This hyperbola has its center at (0, 0) and its transverse axis is the line $y = x$. The asymptotes are the x- and y-axes. Its vertices are at $\left(\sqrt{16}, \sqrt{16}\right) = (4, 4)$ and $\left(-\sqrt{16}, -\sqrt{16}\right) = (-4, -4)$. The graph of this hyperbola is shown in the following figure. To visualize what the graph of the hyperbola will look like, it is often helpful to create an xy-table of the values to produce points on the graph of the equation.

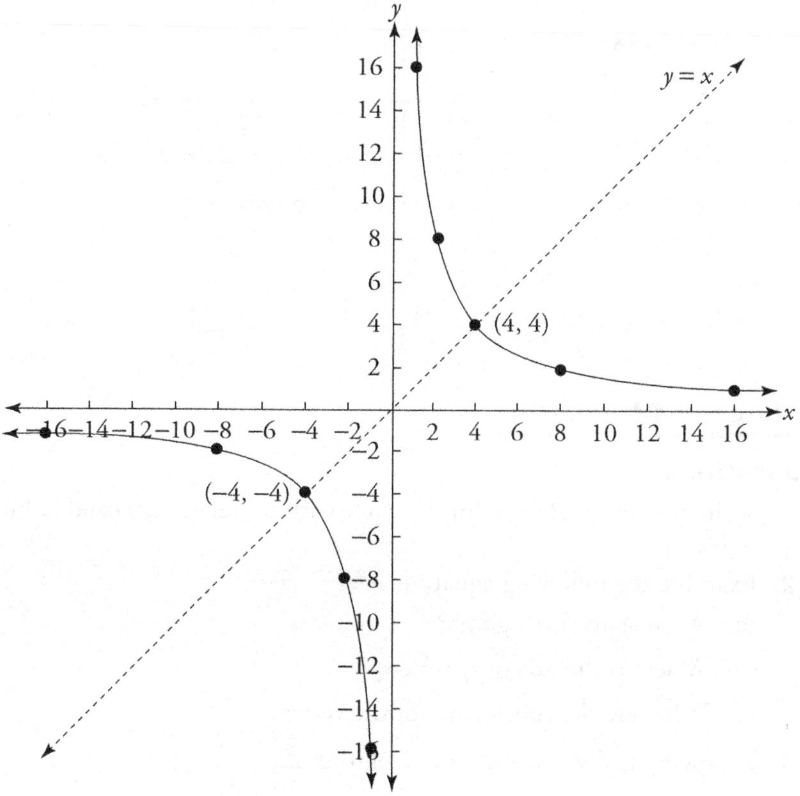

Example 4: Graph $xy = -4$.

This hyperbola has its center at (0, 0). The transverse axis is the line $y = -x$. The asymptotes are the x- and y-axes. The vertices are at $\left(\sqrt{4}, -\sqrt{4}\right) = (2, -2)$ and $\left(-\sqrt{4}, \sqrt{4}\right) = (-2, 2)$.

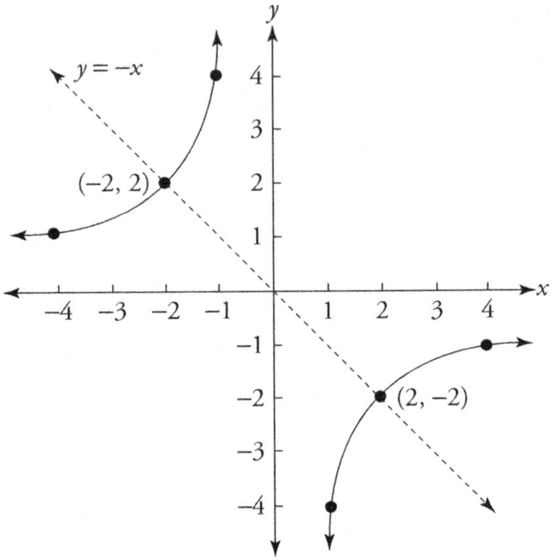

Chapter Check-Out

Questions

1. Write the standard form for the circle with center at (5, 3) and radius $2\sqrt{5}$.
2. Examine the following equation: $x^2 - x - 6 = y$.
   (a) What shape is its graph?
   (b) Where is the axis of symmetry?
   (c) What are the coordinates of the vertex?
3. Following is the equation for an ellipse: $\frac{x^2}{16} + \frac{y^2}{36} = 1$.
   (a) Is its major axis horizontal or vertical?
   (b) What are the coordinates of its foci?
   (c) What is the length of the major axis?
   (d) What is the length of the minor axis?

4. How does the ellipse differ from the circle?
5. Compare and contrast the ellipse and the hyperbola.
6. Find two numbers whose difference is 20, and whose product is a minimum. Explain what is meant here by "minimum."

Answers

1. $(x - 5)^2 + (y - 3)^2 = 20$
2. **(a)** parabola; **(b)** $x = \frac{1}{2}$; **(c)** $\left(\frac{1}{2}, -6\frac{1}{4}\right)$
3. **(a)** vertical; **(b)** $\left(0, 2\sqrt{5}\right)$ and $\left(0, -2\sqrt{5}\right)$; **(c)** 12; **(d)** 8
4. Answers may vary. The standard form of a circle does *not* have fractions and does *not* = 1 unless the radius itself = 1.
5. Answers may vary. Standard form of their equations appear to be similar, with ellipses being sums and hyperbolas being differences. Graphing an ellipse depends on the larger denominator, whereas graphing a hyperbola depends on the first appearing variable.
6. The numbers are 10 and −10. The minimum product is −100.

   The numbers are x and $(x - 20)$.

   The expression of their product is $x\,(x - 20)$.

   The word "minimum" implies finding the coordinates of the vertex of the parabola.

$$y = x^2 - 20x$$

$$\left(-\frac{b}{2a},\ \right)$$

$$\left(\overbrace{10,}^{\text{First Number}}\quad \overbrace{-100}^{\text{Product}}\right)$$

The second number is:

$$x - 20 =$$

$$10 - 20 = -10$$

This is further proof that −100 is the product.

Answers

Chapter 13
QUADRATIC SYSTEMS

Chapter Check-In

- ❑ Solving systems of equations algebraically
- ❑ Solving systems of equations graphically
- ❑ Solving systems of inequalities graphically

Common Core Standard: Linear, Quadratic, and Exponential Models

Represent and solve linear equations and inequalities in two variables algebraically and graphically in quadratic systems (A.REI.7.11). Analyze functions using different representations (F.IF.7).

Earlier chapters explained systems of linear equations and explored various strategies for solving them. This chapter discusses systems containing quadratic equations and strategies for solving them. Solving a system of linear equations means finding the points of intersection of the given lines, if they intersect. When solving a system involving quadratic equations, it is helpful to identify the shapes; this is based on knowing the standard forms of their equations. Then you can visualize what the outcome may look like and, therefore, be conscious of how many solutions to expect to find for the given system. This chapter also deals with solving quadratic systems of inequalities by graphing. The end result will be a shaded region.

Systems of Equations Solved Algebraically

When given two equations in two variables, there are essentially two algebraic methods for solving them. One is substitution and the other is elimination.

Example 1: Solve the following system of equations algebraically.

(1) $x^2 + 2y^2 = 10$
(2) $3x^2 - y^2 = 9$

Recognize this system as the intersection of an ellipse and a hyperbola; thus, you can expect up to four points of intersection.

This system is more easily solved using the elimination method.

equation (1)	$x^2 + 2y^2 = 10$
2 times equation (2)	$6x^2 - 2y^2 = 18$
Add the results	$7x^2 = 28$
	$x^2 = 4$
	$x = \pm 2$

Using equation (1), substitute the two possible values for x to find the solution for this system.

When $x = 2$	When $x = -2$
$x^2 + 2y^2 = 10$	$x^2 + 2y^2 = 10$
$(2)^2 + 2y^2 = 10$	$(-2)^2 + 2y^2 = 10$
$4 + 2y^2 = 10$	$4 + 2y^2 = 10$
$2y^2 = 6$	$2y^2 = 6$
$y^2 = 3$	$y^2 = 3$
$y = \pm\sqrt{3}$	$y = \pm\sqrt{3}$
$\left(2, \sqrt{3}\right), \left(2, -\sqrt{3}\right)$	$\left(-2, \sqrt{3}\right), \left(-2, -\sqrt{3}\right)$

The solution consists of the four ordered pairs.

Example 2: Solve the following system of equations algebraically.

(1) $x^2 + y^2 = 100$
(2) $x - y = 2$

Recognize this to be the intersection of a circle and a line; thus, you can expect up to two points of intersection.

This system is more easily solved using substitution. Solve equation (2) for x, then substitute that result for x in equation (1).

Solving equation (2) for x,

$$x - y = 2$$
$$x = y + 2$$

Substituting into equation (1),

$$x^2 + y^2 = 100$$
$$(y+2)^2 + y^2 = 100$$
$$y^2 + 4y + 4 + y^2 = 100$$
$$2y^2 + 4y + 4 = 100$$
$$2y^2 + 4y - 96 = 0$$
$$2(y^2 + 2y - 48) = 0$$
$$2(y-6)(y+8) = 0$$
$$y - 6 = 0 \text{ or } y + 8 = 0$$
$$y = 6 \text{ or } y = -8$$

Using equation (2), substitute the two possible values for y to find the solution for this system. ***Note:*** You may want to find the second component of the point using the less time-consuming equation so you can avoid finding extraneous roots.

When $y = 6$	When $y = -8$
$x - y = 2$	$x - y = 2$
$x - 6 = 2$	$x - (-8) = 2$
$x = 8$	$x = -6$
$(8, 6)$	$(-6, -8)$

The solution consists of the two ordered pairs.

Systems of Equations Solved Graphically

Graphs can be used to solve systems containing quadratic equations. This method, however, often allows only approximate solutions, whereas the algebraic method arrives at exact solutions. We will now solve the same two previous systems graphically.

Example 1: Solve the following system of equations graphically.

(1) $x^2 + 2y^2 = 10$

(2) $3x^2 - y^2 = 9$

Equation (1) is the equation of an ellipse. Convert the equation into standard form.

$$x^2+2y^2=10$$

$$\frac{x^2}{10}+\frac{2y^2}{10}=\frac{10}{10}$$

$$\frac{x^2}{10}+\frac{y^2}{5}=1$$

The major intercepts are at $\left(\sqrt{10},\ 0\right)$ and $\left(-\sqrt{10},\ 0\right)$, and the minor intercepts are at $\left(0,\ \sqrt{5}\right)$ and $\left(0,\ -\sqrt{5}\right)$.

Equation (2) is the equation of a hyperbola. Convert the equation into standard form.

$$3x^2-y^2=9$$

$$\frac{3x^2}{9}-\frac{y^2}{9}=\frac{9}{9}$$

$$\frac{\overset{1}{\cancel{3}}x^2}{\underset{3}{\cancel{9}}}-\frac{y^2}{9}=\frac{9}{9}$$

$$\frac{x^2}{3}-\frac{y^2}{9}=1$$

The transverse axis is horizontal, and the vertices are at $\left(\sqrt{3},\ 0\right)$ and $\left(-\sqrt{3},\ 0\right)$, as shown in the following figure, which depicts approximate solutions to the hyperbola and ellipse.

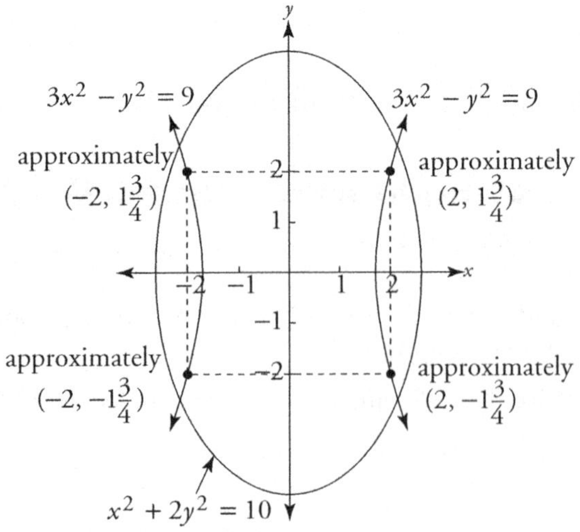

The approximate answers are

$$\left\{\left(-2, 1\frac{3}{4}\right), \left(-2, -1\frac{3}{4}\right), \left(2, 1\frac{3}{4}\right), \left(2, -1\frac{3}{4}\right)\right\}$$

The exact answers, as found in Example 1 in "Systems of Equations Solved Algebraically" (p. 242) are

$$\left\{\left(-2, \sqrt{3}\right), \left(-2, -\sqrt{3}\right), \left(2, \sqrt{3}\right), \left(2, -\sqrt{3}\right)\right\}$$

Example 2: Solve the following system of equations graphically.

(1) $x^2 + y^2 = 100$

(2) $x - y = 2$

Equation (1) is the equation of a circle centered at (0, 0) with a radius of 10. Equation (2) is the equation of a line. The solutions are

$$\{(-6, -8), (8, 6)\}$$

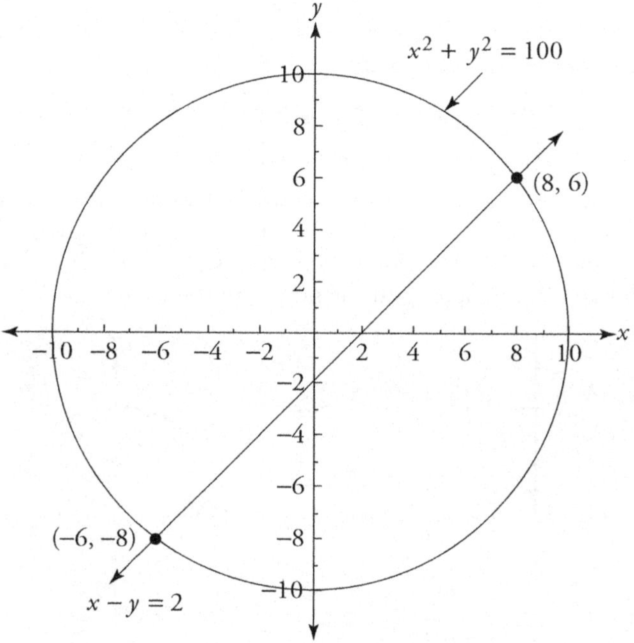

Compare the solution to the same solution found in Example 2 in "Systems of Equations Solved Algebraically" (pp. 242–243).

Systems of Inequalities Solved Graphically

To graph the solution of a system of inequalities, graph each inequality and find the overlap of the two shaded regions.

Example 1: Graph the solution for the following system.

(1) $x^2 + y^2 \le 16$

(2) $y \le x^2 + 2$

Equation (1) is the equation of a circle centered at (0, 0) with a radius of 4. Graph the circle, then select a test point not on the circle and place it into the original inequality. If that result is true, then shade the region where the test point is located. Otherwise, shade the other region. Use (0, 0) as a test point.

$$x^2 + y^2 \le 16$$

$$0^2 + 0^2 \le 16 \checkmark$$

This is a true statement. Therefore, the interior of the circle is shaded (in the figure, this shading is done with horizontal lines).

Equation (2) is the equation of a parabola opening upward with its vertex at (0, 2). Use (0, 0) as a test point.

$$y \le x^2 + 2$$

$$0 \le 0^2 + 2 \checkmark$$

This is a true statement. Therefore, shade the exterior of the parabola. In the figure, this shading is done with vertical lines. The region with both shadings represents the solution of the systems of inequalities. That solution is shown by the shading in the graph on the right.

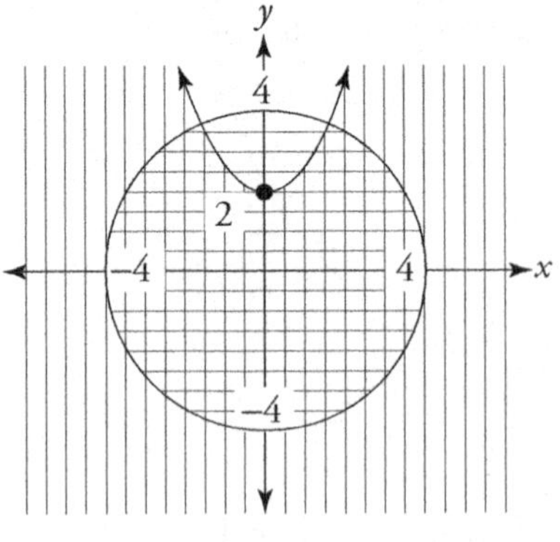

(a)

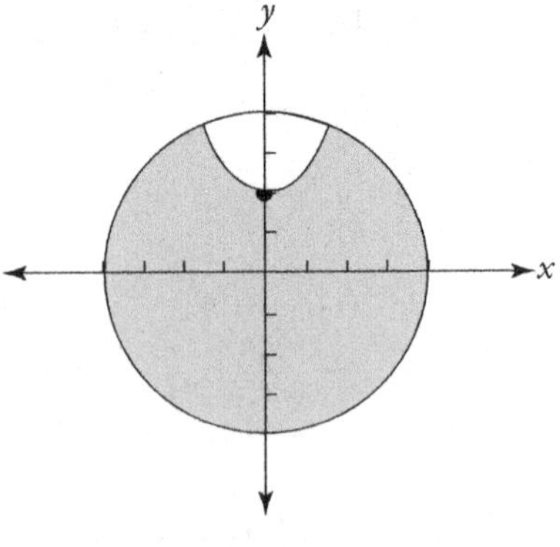

(b)

Example 2: Solve the following system of inequalities graphically.

(1) $\frac{x^2}{36}+\frac{y^2}{25}\le 1$

(2) $\frac{y^2}{4}-\frac{x^2}{1}>1$

Equation (1) is the equation of an ellipse centered at (0, 0) with major intercepts at (6, 0) and (−6, 0) and minor intercepts at (0, 5) and (0, −5). Use (0, 0) as a test point.

$$\frac{x^2}{36}+\frac{y^2}{25}\le 1$$

$$\frac{0^2}{36}+\frac{0^2}{25}\le 1 \checkmark$$

This is a true statement. Therefore, shade the interior of the ellipse (in the figure, this shading is done with horizontal lines).

Equation (2) is the equation of a hyperbola centered at (0, 0) opening vertically with vertices at (0, 2) and (0, −2). Use (0, 0) as a test point.

$$\frac{y^2}{4}-\frac{x^2}{1}>1$$

$$\frac{0^2}{4}-\frac{0^2}{1}\not>1$$

This is not a true statement. Therefore, shade the area inside the curves of the hyperbola (in the figure, this shading is done with vertical lines). The region with both shadings represents the solution to the system of inequalities. That solution is shown in the graph on the right.

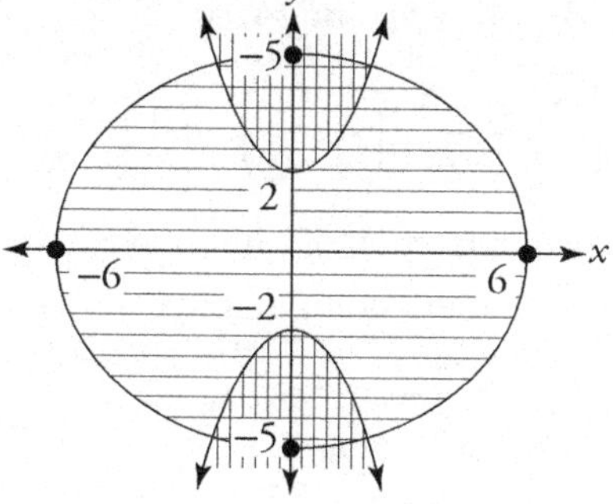

(a)

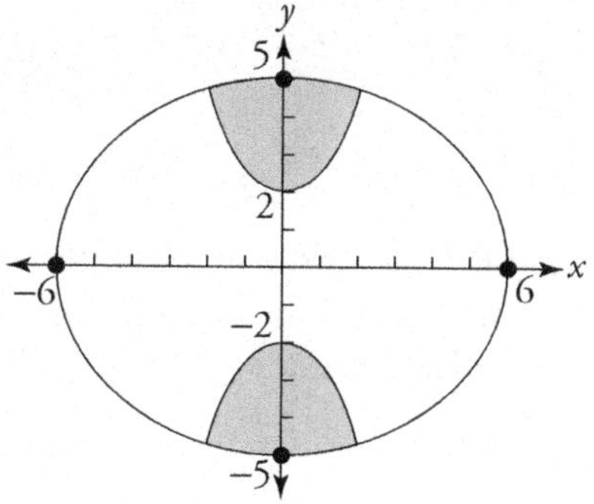

(b)

Chapter Check-Out

Questions

1. Solve by substitution:

   (1) $x^2 + 3y^2 = 172$

   (2) $2x - y = 3$

2. Solve by elimination:

   (1) $2x^2 - 3y^2 = 5$

   (2) $3x^2 + 2y^2 = 66$

3. Solve the following system using any method.

   (1) $16x^2 + 4y^2 = 10$

   (2) $8y^2 - 32x^2 = -16$

4. Predict the number of solutions to the following systems based on the intersection of their shapes:

   (a) (1) $y = -x - 2$

   (2) $x^2 + y^2 = 16$

   (b) (1) $x^2 + y^2 = 49$

   (2) $\frac{x^2}{16} - \frac{y^2}{9} = 1$

5. Graph the following system of inequalities:

   (1) $x^2 + y^2 < 25$

   (2) $y < -\frac{1}{4}x^2 + 6$

Answers

1. $x = 5, y = 7; \left(-\frac{29}{13}, -\frac{97}{13}\right)$
2. $x = \pm 4, y = \pm 3$; $(-4, -3)$, $(-4, 3)$, $(4, -3)$, and $(4, 3)$
3. $x = \pm\frac{3}{4}, y = \pm\frac{1}{2}; \left(-\frac{3}{4}, -\frac{1}{2}\right), \left(\frac{3}{4}, -\frac{1}{2}\right), \left(\frac{3}{4}, \frac{1}{2}\right)$, and $\left(-\frac{3}{4}, \frac{1}{2}\right)$
4. **(a)** Up to two solutions since it is the intersection of a line and a circle.

   (b) Up to four solutions since it is the intersection of a hyperbola and a circle.

5. The solution is the overlap of the region inside the circle of radius 5 centered at the origin, and the region inside the vertically downward parabola whose vertex is (0, 6).

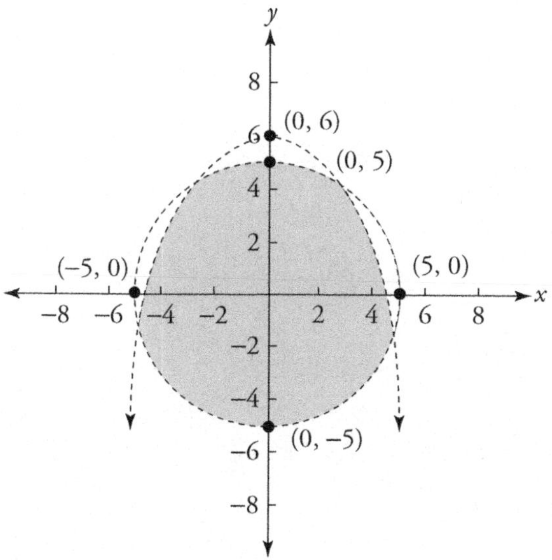

Chapter 14

EXPONENTIAL AND LOGARITHMIC FUNCTIONS

Chapter Check-In

- ❑ Defining exponential functions and their inverses
- ❑ Examining logarithmic functions
- ❑ Examining properties of logarithms
- ❑ Solving exponential and logarithmic equations
- ❑ Applying exponential and log functions

Common Core Standard: Linear, Quadratic, and Exponential Models

Represent and solve linear equations and inequalities in two variables algebraically and graphically in quadratic systems (A.REI.7.11). Analyze functions using different representations (F.IF.7).

Exponential functions can increase or decrease (functions *increase* or *decrease*; graphs *rise* and *fall*) very, very quickly. The following list of the powers of 2 represents the function $f(x)=2^x$. As x increases in small amounts, the function increases by far greater amounts.

$$\begin{array}{lll} 2^0=1 & 2^4=16 & 2^8=256 \\ 2^1=2 & 2^5=32 & 2^9=512 \\ 2^2=4 & 2^6=64 & 2^{10}=1{,}024 \\ 2^3=8 & 2^7=128 & 2^{11}=2{,}048 \end{array}$$

Logarithmic functions are the inverses of exponential functions. This chapter introduces the concept of logarithms and how any exponential function can be expressed in logarithmic form. Similarly, all logarithmic functions can be rewritten in exponential form. Logarithms are useful when working with very large numbers by allowing us to manipulate numbers of a much more manageable size. Although logarithms of base 10 and base e are those most often dealt with in mathematics, this chapter shows how to change logs with other bases to logs of base 10 and base e.

Exponential Functions

Any function defined by $y = b^x$, where $b > 0$, $b \neq 1$, and x is a real number, is called an **exponential function.**

Example 1: Graph $y = 2^x$, or $f(x) = 2^x$.

First find a sufficient number of ordered pairs to see the shape of the graph.

x	$2^x = y$	(x, y) $(x, f(x))$
−3	$2^{-3} = \frac{1}{8}$	$\left(-3, \frac{1}{8}\right)$
−2	$2^{-2} = \frac{1}{4}$	$\left(-2, \frac{1}{4}\right)$
−1	$2^{-1} = \frac{1}{2}$	$\left(-1, \frac{1}{2}\right)$
0	$2^0 = 1$	(0, 1)
1	$2^1 = 2$	(1, 2)
2	$2^2 = 4$	(2, 4)
3	$2^3 = 8$	(3, 8)

Plot these points and connect them to form a smooth curve. No value of x will make y become zero. The more negative x becomes, the smaller y becomes. The negative x-axis becomes an **asymptote line** for this function, as shown in the following figure.

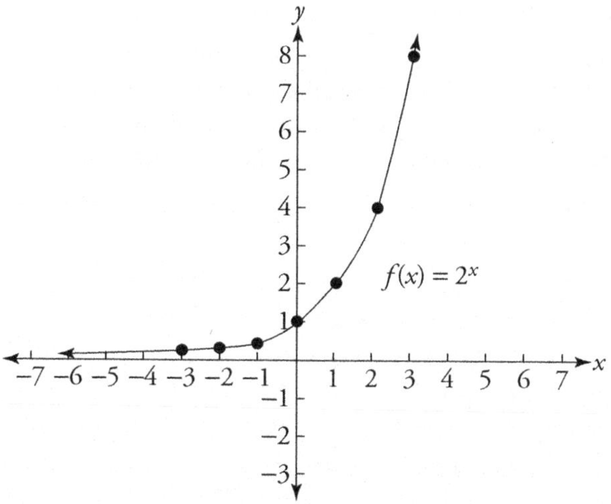

Example 2: Graph $f(x) = \left(\frac{1}{2}\right)^x$; this is function form for $y = \left(\frac{1}{2}\right)^x$.

Make a chart and plot the ordered pairs.

x	$\left(\frac{1}{2}\right)^x = f(x)$	***[x, f(x)]***
−3	$\left(\frac{1}{2}\right)^{-3} = 8$	(−3, 8)
−2	$\left(\frac{1}{2}\right)^{-2} = 4$	(−2, 4)
−1	$\left(\frac{1}{2}\right)^{-1} = 2$	(−1, 2)
0	$\left(\frac{1}{2}\right)^{0} = 1$	(0, 1)
1	$\left(\frac{1}{2}\right)^{1} = \frac{1}{2}$	$\left(1, \frac{1}{2}\right)$
2	$\left(\frac{1}{2}\right)^{2} = \frac{1}{4}$	$\left(2, \frac{1}{4}\right)$
3	$\left(\frac{1}{2}\right)^{3} = \frac{1}{8}$	$\left(3, \frac{1}{8}\right)$

The graph is shown in the figure below.

All exponential functions, $f(x) = b^x$, $b > 0$, $b \neq 1$, will contain the ordered pair (0, 1), since $b^0 = 1$ for all $b \neq 0$. Exponential functions with $b > 1$ will have a basic shape like that in the graph shown in Example 1, and exponential functions with $b < 1$ will have a basic shape like that of the graph below.

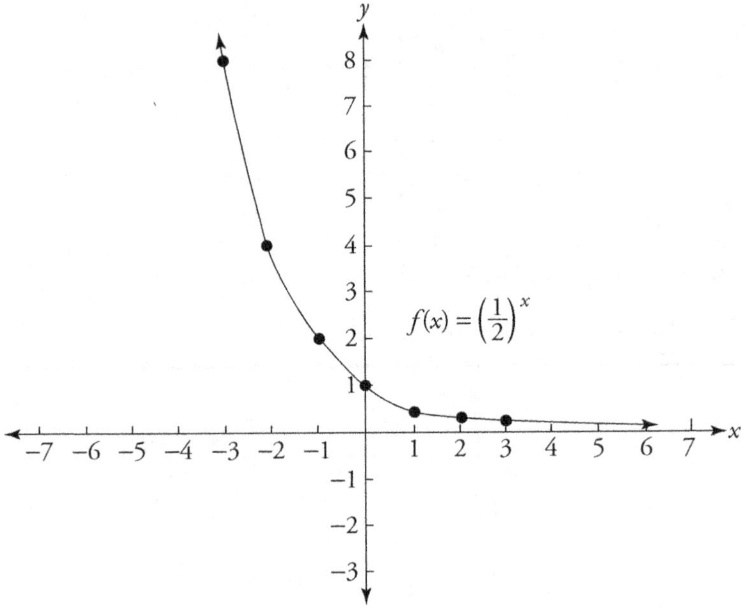

The graph of $x = b^y$ is called the *inverse* of the graph of $y = b^x$ because the x and y variables are interchanged. The graphs of inverses are symmetrical about the line $y = x$. That is, if the graph of $y = b^x$ is "folded over" the line $y = x$ and then retraced, it creates the graph of $x = b^y$. Whatever ordered pairs satisfy $y = b^x$, the reversed ordered pairs would satisfy $x = b^y$.

Example 3: Graph $y = 3^x$ and $x = 3^y$ on the same set of axes.

x	$3^x = y$	(x, y)
-3	$(3)^{-3} = \frac{1}{27}$	$\left(-3, \frac{1}{27}\right)$
-2	$(3)^{-2} = \frac{1}{9}$	$\left(-2, \frac{1}{9}\right)$
-1	$(3)^{-1} = \frac{1}{3}$	$\left(-1, \frac{1}{3}\right)$
0	$(3)^0 = 1$	(0, 1)
1	$(3)^1 = 3$	(1, 3)
2	$(3)^2 = 9$	(2, 9)
3	$(3)^3 = 27$	(3, 27)

x	$3^y = x$	(x, y)
$\frac{1}{27}$	$(3)^{-3} = \frac{1}{27}$	$\left(\frac{1}{27}, -3\right)$
$\frac{1}{9}$	$(3)^{-2} = \frac{1}{9}$	$\left(\frac{1}{9}, -2\right)$
$\frac{1}{3}$	$(3)^{-1} = \frac{1}{3}$	$\left(\frac{1}{3}, -1\right)$
1	$(3)^0 = 1$	(1, 0)
3	$(3)^1 = 3$	(3, 1)
9	$(3)^2 = 9$	(9, 2)
27	$(3)^3 = 27$	(27, 3)

The graphs of $y = 3^x$ and $x = 3^y$ are shown below.

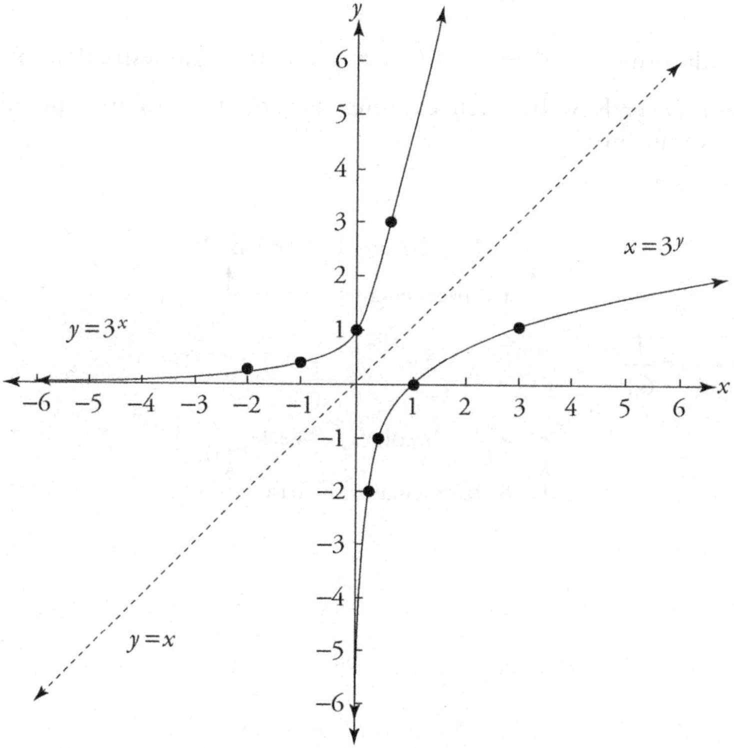

Logarithmic Functions

If $x = 2^y$ were to be solved for y so that it could be written in function form, a new word or symbol would need to be introduced. The word **logarithm,** abbreviated **log,** is introduced to satisfy this need.

This equation is written as $y = \log_2 x$.

Notice that the 2 is the base in either form, exponential or logarithmic.

This is read as "y equals the log, base 2, of x."

A **logarithmic function** is a function of the form

$$y = \log_b x \qquad x > 0, \text{ where } b > 0 \text{ and } b \neq 1$$

which is read "y equals the log, base b, of x."

$$y = \log_b x \quad \text{is equivalent to} \quad x = b^y$$

the base remains the base

In both forms, $x > 0$ and $b > 0$, $b \neq 1$. There are no restrictions on y.

Example 1: Rewrite each exponential equation in its equivalent logarithmic form.

(a) $5^2 = 25$

$$5^2 = 25 \quad \text{becomes} \quad 2 = \log_5 25$$

the base remains the base

(b) $4^{-3} = \dfrac{1}{64}$

$$4^{-3} = \frac{1}{64} \quad \text{becomes} \quad -3 = \log_4 \left(\frac{1}{64}\right)$$

the base remains the base

(c) $\left(\frac{1}{2}\right)^{-4} = 16$

$$\left(\frac{1}{2}\right)^{-4} = 16 \quad \text{becomes} \quad -4 = \log_{(1/2)}16$$

the base remains the base

Example 2: Rewrite each logarithmic equation in its equivalent exponential form.

(a) $\log_6 36 = 2$

$$\log_6 36 \quad \text{becomes} \quad 6^2 = 36$$

base

(b) $\log_a m = p$

$$\log_a m = p \quad \text{becomes} \quad a^p = m$$

base

Example 3: Solve each of the following equations, if possible.

(a) $\log_7 49 = y$ becomes $7^y = 49$

$$7^y = 7^2$$

$$y = 2$$

(b) $\log_2\left(\frac{1}{8}\right) = y$ becomes $2^y = \frac{1}{8}$

$$2^y = 2^{-3}$$

$$y = -3$$

(c) $\log_y 8 = 3$ becomes $y^3 = 8$

$$y^3 = 2^3$$

$$y = 2$$

(d) $\log_4 y = -2$ becomes $4^{-2} = y$

$$\frac{1}{16} = y$$

(e) $\log_3(-9) = y$ becomes $3^y = -9$

The equation in item (e) cannot be solved since 3^y will always be a positive result. The log of a negative number does not exist. Base numbers are always positive; exponents can be either positive or negative. A positive number raised to a positive power is still positive. A positive number raised to a negative number is also positive. For example,

$$2^3 = 8$$

$$2^{-3} = \frac{1}{8}$$

Therefore,

$$\log_2(8) = 3$$

$$\log_2\left(\frac{1}{8}\right) = -3$$

The bases used most often when working with logarithms are base 10 and base *e*. (The letter *e* represents an irrational number that has many applications in mathematics and science. The value of *e* is approximately 2.718281828....) Log_{10} is known as the **common logarithm** and is written as log; when the base is not written, it is understood to be 10. Log_e is known as the **natural logarithm** and is written as ln.

Example 4: Find the following logarithms.

(a) log 100

$$\text{Let } \log 100 = x$$

↑base 10 is understood

$$10^x = 100$$

$$10^x = 10^2$$

$$x = 2$$

$$\log 100 = 2$$

(b) log 10,000

$$\text{Let } \log 10{,}000 = x$$

$$10^x = 10{,}000$$

$$10^x = 10^4$$

$$x = 4$$

$$\log 10{,}000 = 4$$

(c) log 0.1

$$\text{Let } \log 0.1 = x$$

$$10^x = 0.1$$

$$10^x = 10^{-1}$$

$$x = -1$$

$$\log 0.1 = -1$$

(d) ln *e*

$$\text{Let } \ln e = x$$

↑base *e* is understood

$$\log_e e = x$$

$$e^x = e$$

$$e^x = e^1$$

$$x = 1$$

$$\ln e = 1$$

(e) $\ln e^2$

$$\begin{aligned} \text{Let } \ln e^2 &= x \\ \log_e e^2 &= x \\ e^x &= e^2 \\ x &= 2 \\ \ln e^2 &= 2 \end{aligned}$$

Properties of Logarithms

The properties of logarithms assume the following about the variables M, N, b, and x: $M > 0$, $N > 0$, $b > 0$, $b \neq 1$, x is a real number. The properties are as follows:

1. $\log_b b = 1$
2. $\log_b 1 = 0$
3. $\log_b b^x = x$
4. $b^{\log_b x} = x$
5. $\log_b(MN) = \log_b(M) + \log_b(N)$
6. $\log_b\left(\dfrac{M}{N}\right) = \log_b M - \log_b N$ (**Note:** Do not confuse $\log_b\left(\dfrac{M}{N}\right)$ with $\dfrac{\log_b M}{\log_b N}$. The latter is the quotient of the two different logs.)
7. $\log_b M^x = x \log_b M$
8. If $\log_b x = \log_b y$, then $x = y$.
9. $\log_b x = \dfrac{\log x}{\log b}$ (**Note:** This is known as the *change of base formula.* Using the formula allows us to perform the calculations in base 10.)

Example 1: The following items are simplified and the respective property is noted.

(a) $\log_7 7 = 1$ (property 1)

(b) $\log_5 1 = 0$ (property 2)

(c) $\log_4 4^3 = 3$ (property 3)

(d) $6^{\log_6 5} = 5$ (property 4)

Example 2: If $\log_3 5 \approx 1.5$, $\log_3 3 = 1$ and $\log_3 2 \approx 0.6$, approximate each of the following by using the properties of logarithms.

(a) $\log_3 10$

$$\begin{aligned}\log_3 10 &= \log_3(5\cdot 2)\\ &= \log_3 5 + \log_3 2 && \text{(property 5)}\\ &\approx 1.5 + 0.6\\ &\approx 2.1\end{aligned}$$

(b) $\log_3\left(\frac{5}{2}\right)$

$$\begin{aligned}\log_3\left(\frac{5}{2}\right) &= \log_3 5 - \log_3 2 && \text{(property 6)}\\ &\approx 1.5 - 0.6\\ &\approx 0.9\end{aligned}$$

(c) $\log_3 25$

$$\begin{aligned}\log_3 25 &= \log_3 5^2\\ &= 2\log_3 5 && \text{(property 7)}\\ &\approx 2(1.5)\\ &\approx 3\end{aligned}$$

(d) $\log_3\sqrt{5}$

$$\begin{aligned}\log_3\sqrt{5} &= \log_3 5^{\frac{1}{2}}\\ &= \frac{1}{2}\log_3 5 && \text{(property 7)}\\ &\approx \frac{1}{2}(1.5)\\ &\approx .75\end{aligned}$$

It is necessary to use a property of exponents in item (d) to change a root to a power.

(e) $\log_3 1.5$

$$\begin{aligned}\log_3 1.5 &= \log_3\left(\frac{3}{2}\right)\\ &= \log_3 3 - \log_3 2 && \text{(property 6)}\\ &\approx 1 - 0.6 && \text{(property 1)}\\ &\approx 0.4\end{aligned}$$

(f) $\log_3 200$

$$\begin{aligned}\log_3 200 &= \log_3[(2^3)(5^2)]\\ &= \log_3 2^3 + \log_3 5^2 && \text{(property 5)}\\ &= 3\log_3 2 + 2\log_3 5 && \text{(property 7)}\\ &\approx 3(0.6) + 2(1.5)\\ &\approx 1.8 + 3\\ &\approx 4.8\end{aligned}$$

Example 3: Rewrite each expression as the logarithm of a single quantity.

(a) $2\log_b x + \frac{1}{3}\log_b y$

$$\begin{aligned}2\log_b x + \frac{1}{3}\log_b y &= \log_b x^2 + \log_b y^{\frac{1}{3}} && \text{(property 7)}\\ &= \log_b x^2 y^{\frac{1}{3}} && \text{(property 5)}\\ &= \log_b x^2 \sqrt[3]{y}\end{aligned}$$

(b) $\frac{1}{2}\log_b(x-2) - \log_b y + 3\log_b z$

$$\begin{aligned}\frac{1}{2}\log_b(x-2) - \log_b y + 3\log_b z &= \log_b (x-2)^{\frac{1}{2}} - \log_b y + \log_b z^3 && \text{(property 7)}\\ &= \log_b \frac{(x-2)^{\frac{1}{2}}}{y} + \log_b z^3 && \text{(property 6)}\\ &= \log_b \frac{z^3(x-2)^{\frac{1}{2}}}{y} && \text{(property 5)}\\ &= \log_b \frac{z^3\sqrt{x-2}}{y} && \text{(property of exponents)}\end{aligned}$$

Exponential and Logarithmic Equations

An **exponential equation** is an equation in which the variable appears in an exponent. A **logarithmic equation** is an equation that involves the logarithm of an expression containing a variable.

Exponential equations

To solve exponential equations, first see if both sides of the equation can be written as powers of the same number. If not, take the common logarithm of both sides of the equation and apply property 7.

Example 1: Solve each of the following equations for x.

(a) $3^x = 81$

$$3^x = 3^4$$
$$x = 4$$

(b)
$$3^x = 5$$
$$\log(3^x) = \log 5$$
$$x \log 3 = \log 5 \quad \text{(property 7)}$$

Dividing both sides by log 3,

$$x = \frac{\log 5}{\log 3}$$

Using a calculator for approximation,

$$x \approx \frac{0.699}{0.477}$$
$$\approx 1.465$$

(c)
$$6^{x-3} = 2$$
$$\log(6^{x-3}) = \log 2$$
$$(x-3)\log 6 = \log 2 \quad \text{(property 7)}$$

Dividing both sides by log 6,

$$x - 3 = \frac{\log 2}{\log 6}$$
$$x = \frac{\log 2}{\log 6} + 3$$

Using a calculator for approximation,

$$x \approx \frac{0.301}{0.778} + 3$$
$$\approx 3.387$$

(d)

$$2^{3x-1} = 3^{2x-2}$$
$$\log(2^{3x-1}) = \log(3^{2x-2})$$
$$(3x-1)\log 2 = (2x-2)\log 3 \quad \text{(property 7)}$$

Using the distributive property,

$$3x \log 2 - \log 2 = 2x \log 3 - 2 \log 3$$

Gathering all terms involving the variable on one side of the equation,

$$3x \log 2 - 2x \log 3 = \log 2 - 2 \log 3$$

Factoring out an x,

$$x(3 \log 2 - 2 \log 3) = \log 2 - 2 \log 3$$

Dividing both sides by $3 \log 2 - 2 \log 3$,

$$x = \frac{\log 2 - 2\log 3}{3\log 2 - 2\log 3}$$
$$= \frac{\log 2 - \log 3^2}{\log 2^3 - \log 3^2}$$
$$= \frac{\log 2 - \log 9}{\log 8 - \log 9}$$
$$= \frac{\log\left(\frac{2}{9}\right)}{\log\left(\frac{8}{9}\right)}$$

Using a calculator for approximation,

$$x \approx 12.770$$

Logarithmic equations

To solve an equation involving logarithms, use the properties of logarithms to write the equation in the form $\log_b M = N$ and then change this to exponential form, $M = b^N$.

Example 2: Solve each of the following equations for x.

(a) $\log_4(3x - 2) = 2$

Change to exponential form.

$$(3x-2)=4^2$$
$$3x-2=16$$
$$3x=18$$
$$x=6$$

Check the answer.

$$\log_4(3x-2)=2$$
$$\log_4[3(6)-2]\stackrel{?}{=}2$$
$$\log_4 16\stackrel{?}{=}2$$
$$4^2=16 \checkmark$$

This is a true statement. Therefore, the solution is $x = 6$.

(b) $\log_3 x + \log_3(x - 6) = 3$

$$\log_3 x+\log_3(x-6)=3$$
$$\log_3[x(x-6)]=3 \quad \text{(property 5)}$$

Change to exponential form.

$$x(x-6)=3^3$$
$$x^2-6x=27$$
$$x^2-6x-27=0$$
$$(x-9)(x+3)=0$$
$$x-9=0 \text{ or } x+3=0$$
$$x=9 \text{ or } x=-3$$

Check the answers.

$$x=9$$
$$\log_3 x+\log_3(x-6)=3$$
$$\log_3 9+\log_3(9-6)\stackrel{?}{=}3$$
$$2+1=3 \checkmark$$

$$x=-3$$
$$\log_3 x+\log_3(x-6)=3$$
$$\log_3(-3)+\log_3(-3-6)\stackrel{?}{=}3 \text{ No}$$

Since the logarithm of a negative number is not defined, the only solution is $x = 9$.

(c) $\log_2 (5 + 2x) - \log_2 (4 - x) = 3$

$$\log_2(5+2x)-\log_2(4-x)=3$$

$$\log_2\left(\frac{5+2x}{4-x}\right)=3 \quad \text{(property 6)}$$

Change to exponential form.

$$\frac{5+2x}{4-x}=2^3$$

$$\frac{5+2x}{4-x}=8$$

Using the cross products rule,

$$5+2x=8(4-x)$$

$$5+2x=32-8x$$

$$10x=27$$

$$x=2.7$$

Check the answer.

$$\log_2(5+2x)-\log_2(4-x)=3$$

$$\log_2[5+2(2.7)]-\log_2(4-2.7)\stackrel{?}{=}3$$

$$\log_2(10.4)-\log_2(1.3)\stackrel{?}{=}3$$

$$\log_2\left(\frac{10.4}{1.3}\right)\stackrel{?}{=}3$$

$$\log_2(8)\stackrel{?}{=}8$$

$$2^3=8 \checkmark$$

This is a true statement. Therefore, the solution is $x = 2.7$.

(d) $\log_5 (7x - 9) = \log_5 (x^2 - x - 29)$

$$\log_5(7x-9)=\log_5(x^2-x-29)$$

$$7x-9=x^2-x-29 \qquad \text{(property 8)}$$

$$0=x^2-8x-20$$

$$0=(x-10)(x+2)$$

$$x-10=0 \text{ or } x+2=0$$

$$x=10 \text{ or } x=-2$$

Check the answers.

If $x = 10$,

$$\log_5(7x-9) = \log_5(x^2 - x - 29)$$

$$\log_5[7(10)-9] \stackrel{?}{=} \log_5[10^2 - 10 - 29]$$

$$\log_5(61) = \log_5(61) \checkmark$$

This is a true statement.

If $x = -2$,

$$\log_5(7x-9) = \log_5(x^2 - x - 29)$$

$$\log_5[7(-2)-9] \stackrel{?}{=} \log_5[(-2)^2 - (-2) - 29]$$

$$\log_5(-23) \stackrel{?}{=} \log_5(-23)$$

This appears to be true, but $\log_5(-23)$ is not defined. Therefore, the only solution is $x = 10$.

Example 3: Find $\log_3 8$.

$$\log_3 8 = \frac{\log 8}{\log 3} \quad \text{(property 9, the change of base formula)}$$

Note: $\log 8 = \log_{10} 8$ and $\log 3 = \log_{10} 3$.

Using a calculator for approximation,

$$\log_3 8 \approx \frac{0.903}{0.477}$$

$$\approx 1.893$$

Applications of Exponential and Log Functions

Use this formula for compound interest.

$$A = P\left[1 + \frac{r}{n}\right]^{nt}$$

P is the investment, r is the interest rate per year, n is the number of times the interest is compounded yearly, t is the number of years of the investment, and A is the amount of money accumulated.

When the time is missing, logs are needed to solve for the t.

Example 1: An original investment of $1,000 has grown to $7,450 in a savings account that is compounded quarterly and pays an annual interest rate of 12%. Find the length of time, in years, it took for that original investment to grow to that amount.

$$A = P\left[1+\frac{r}{n}\right]^{nt}$$

$$7{,}450 = 1{,}000\left[1+\frac{0.12}{4}\right]^{4t}$$

$$7.45 = (1+.03)^{4t}$$

$$\log 7.45 = 4t \log 1.03$$

$$t = \frac{\log 7.45}{4\log 1.03}$$

$$t \approx \frac{.8722}{4(.0128)}$$

$$t \approx 17.0$$

It would take about 17 years for the original investment to grow to the new amount.

Chapter Check-Out

Questions

1. Rewrite each equation in logarithmic form.

   (a) $6^2 = 36$ (b) $3^{-3} = \frac{1}{27}$ (c) $\left(\frac{1}{5}\right)^{-3} = 125$

2. Solve the following equations for x, if possible.

   (a) $\log_9 81 = x$ (b) $\log_2\left(\frac{1}{16}\right) = x$ (c) $\log_9 x = -3$

3. Find the following common logarithms:

   (a) log 1,000 (b) log .01 (c) log 100,000

4. Explain the relationship between an exponential function and its log function.
5. Solve for x:

   (a) $2.7^x = 52.3$ (b) $7^x = 343$ (c) $2^{3x} = 3^{x+1}$

6. How long, in years, will it take a $2,500 investment to triple if it is invested at 6% and is compounded quarterly?

Answers

1. (a) $2 = \log_6 36$; (b) $-3 = \log_3\left(\frac{1}{27}\right)$; (c) $-3 = \log_{\frac{1}{5}} 125$
2. (a) $x = 2$; (b) $x = -4$; (c) $x = \frac{1}{729}$
3. (a) 3; (b) -2; (c) 5
4. Answers may vary. The two functions are inverses of each other.
5. (a) $x \approx 3.9839$; (b) $x = 3$; (c) $x \approx 1.1202$
6. 18.4 years

Chapter 15
SEQUENCES AND SERIES

Chapter Check-In

- ❑ Defining and using arithmetic sequences
- ❑ Defining and using arithmetic series
- ❑ Defining and using geometric sequences
- ❑ Defining and using geometric series
- ❑ Interpreting summation notation
- ❑ Applying the concepts of sequences and series

Common Core Standard: Arithmetic with Polynomials and Rational Expressions

Define and use sequences and series: arithmetic, geometric, and Σ. Use polynomial identities to solve problems (A.APR.5). Write expressions in equivalent forms to solve problems (A.SSE.4).

Sequences are forms of patterns or progressions. Arithmetic sequences progress, or grow, by common amounts. For example, 1, 5, 9, 13, 17, 21, … is an arithmetic sequence. In this example, each term is four more than the one before it. Geometric sequences involve multiplication by the same number. For example, 3, 12, 48, 192, … is a geometric sequence. In this example, each term is four times the one before it. Geometric sequences grow and diminish much more rapidly than arithmetic sequences.

Series are sums of sequences with a finite or an infinite number of terms. In this chapter, arithmetic and geometric series and sequences, as well as summation notation, are defined and studied.

Definition and Examples of Sequences

A **sequence** is an ordered list of numbers.

$$\left.\begin{array}{l} 1, 3, 5, 7, 9, \ldots \\ -8, 3, 14, 25, \ldots \\ 1, 2, 4, 8, 16, \ldots \end{array}\right\} \text{are examples of sequences}$$

The three dots mean to continue forward in the pattern established. Each number in the sequence is called a **term.** In the sequence 1, 3, 5, 7, 9, …, 1 is the first term, 3 is the second term, 5 is the third term, and so on. The notation $a_1, a_2, a_3, \ldots a_n$ is used to denote the different terms in a sequence. The expression a_n is referred to as the **general** or ***n*th term** of the sequence. We can find any term in a sequence by using the formula for its general term.

Example 1: Find the first five terms of the sequence whose general term is $a_n = 3n + 2$.

$$\begin{aligned} a_n &= 3n + 2 \\ a_1 &= 3(1) + 2 = 5 \\ a_2 &= 3(2) + 2 = 8 \\ a_3 &= 3(3) + 2 = 11 \\ a_4 &= 3(4) + 2 = 14 \\ a_5 &= 3(5) + 2 = 17 \end{aligned}$$

The first five terms are 5, 8, 11, 14, and 17.

Example 2: Find the first five terms of the sequence whose general term is $a_n = 2(3^{n-1})$.

$$\begin{aligned} a_n &= 2(3^{n-1}) \\ a_1 &= 2(3^{1-1}) = 2(3^0) = 2(1) = 2 \\ a_2 &= 2(3^{2-1}) = 2(3^1) = 2(3) = 6 \\ a_3 &= 2(3^{3-1}) = 2(3^2) = 2(9) = 18 \\ a_4 &= 2(3^{4-1}) = 2(3^3) = 2(27) = 54 \\ a_5 &= 2(3^{5-1}) = 2(3^4) = 2(81) = 162 \end{aligned}$$

The first five terms are 2, 6, 18, 54, and 162.

Conversely, we can find the general term when given a sequence.

Example 3: Find an expression for the *n*th term of each sequence.

(a) 2, 4, 6, 8, ...

$$a_1 = 2 = 2(1)$$
$$a_2 = 4 = 2(2)$$
$$a_3 = 6 = 2(3)$$
$$a_4 = 8 = 2(4)$$

Based on this pattern, $a_n = 2n$, this is the formula for its general term.

(b) 10, 50, 250, 1,250, ...

$$a_1 = 10 = 2(5) = 2(5^1)$$
$$a_2 = 50 = 2(25) = 2(5^2)$$
$$a_3 = 250 = 2(125) = 2(5^3)$$
$$a_4 = 1{,}250 = 2(625) = 2(5^4)$$

Based on this pattern, $a_n = 2(5^n)$, this is the formula for its general term.

(c) 3, 7, 11, 15, 19, ...

$$a_1 = 3$$
$$a_2 = 7 = 3 + 4(1)$$
$$a_3 = 11 = 3 + 8 = 3 + 4(2)$$
$$a_4 = 15 = 3 + 12 = 3 + 4(3)$$
$$a_5 = 19 = 3 + 16 = 3 + 4(4)$$

Based on this pattern,

$$a_n = 3 + 4(n - 1)$$
$$a_n = 3 + 4n - 4$$
$$a_n = 4n - 1$$

This is the formula for its general term.

Arithmetic Sequence

An **arithmetic sequence** is a sequence in which each term is found by adding the same value to the previous term. The general term is described by

$$a_n = a_1 + (n - 1)d$$

In this formula, d is called the **common difference.** It is found by taking any term in the sequence and subtracting its preceding term.

Example 1: In each arithmetic sequence, find the common difference, the general term of the sequence in the form $a_n = a_1 + (n - 1)d$, and the twentieth term of the sequence.

(a) 1, 5, 9, 13, 17, ...

The common difference is

$$d = a_2 - a_1 = 5 - 1 = 4$$

The general term of the sequence is

$$a_n = 1 + (n-1)4$$

$$a_n = 4n - 3$$

The twentieth term of the sequence is

$$a_{20} = 4(20) - 3 = 77$$

(b) $-\frac{5}{8}, -\frac{3}{8}, -\frac{1}{8}, \frac{1}{8}, \frac{3}{8}, \ldots$

The common difference is

$$d = a_2 - a_1 = -\frac{3}{8} - \left(-\frac{5}{8}\right) = \frac{1}{4}$$

The general term of the sequence is

$$a_n = -\frac{5}{8} + (n-1)\left(\frac{1}{4}\right)$$

$$a_n = \frac{1}{4}n - \frac{7}{8}$$

The twentieth term of the sequence is

$$a_{20} = \frac{1}{4}(20) - \frac{7}{8} = 5 - \frac{7}{8} = \frac{33}{8}$$

Arithmetic Series

An **arithmetic series** is the sum of the terms in an arithmetic sequence with a finite number of terms. Following is a formula for finding the sum:

Formula 1:

If S_n represents the sum of a finite arithmetic sequence with terms a_1, a_2, a_3, ... a_n, then

$$S_n = \frac{n}{2}(a_1 + a_n)$$

This formula requires knowing the values of the first and last terms and the number of terms.

To study further,

$$a_n = a_1 + (n-1)d$$
$$a_1 + a_n = a_1 + [a_1 + (n-1)d]$$
$$a_1 + a_n = 2a_1 + (n-1)d$$

Substituting this last expression for $(a_1 + a_n)$ into Formula 1, another formula for the sum of an arithmetic sequence is formed.

Formula 2:

$$S_n = \frac{n}{2}[2a_1 + (n-1)d]$$

This formula for the sum of a finite arithmetic sequence requires knowing the first term, the common difference, and the number of terms.

Example 1: In the arithmetic sequence −3, 4, 11, 18, ..., find the sum of the first 20 terms.

First, find the values of a_1, d, and n:

$$a_1 = -3$$
$$d = 4 - (-3) = 7$$
$$n = 20$$

Use Formula 2 to find the sum.

$$\begin{aligned} S_{20} &= \frac{n}{2}[2a_1 + (n-1)d] \\ &= \frac{20}{2}[2(-3) + (20-1)(7)] \\ &= 10(-6+133) \\ &= 1,270 \end{aligned}$$

Example 2: Find the sum of the multiples of 3 between 28 and 112.

The first multiple of 3 between 28 and 112 is 30, and the last multiple of 3 between 28 and 112 is 111. In order to use Formula 1, the number of terms must be known. $a_n = a_1 + (n - 1)d$ can be used to find n.

$$a_n = 111, \quad a_1 = 30, \quad d = 3$$

$$\begin{aligned} 111 &= 30 + (n-1)(3) \\ 81 &= (n-1)(3) \\ 27 &= (n-1) \\ 28 &= n \end{aligned}$$

Now, use Formula 1.

$$\begin{aligned} S_n &= \frac{n}{2}(a_1 + a_n) \\ S_{20} &= \frac{28}{2}(30+111) \\ &= 14(141) \\ &= 1,974 \end{aligned}$$

The sum of the multiples of 3 between 28 and 112 is 1,974.

Geometric Sequence

A **geometric sequence** is a sequence in which each term is found by multiplying the preceding term by the same value. Its general term is

$$a_n = a_1 r^{n-1}$$

The value r is called the **common ratio.** It is found by taking any term in the sequence and dividing it by its preceding term.

Example 1: In each geometric sequence, find the common ratio, the general term in the form $a_n = a_1 r^{n-1}$, and the eighth term of the sequence.

(a) 1, 3, 9, 27, ...

The common ratio is

$$r = \frac{a_2}{a_1} = \frac{3}{1} = 3$$

The general term is

$$a_n = 1(3)^{n-1}$$

The eighth term of the sequence is

$$\begin{aligned} a_8 &= 1(3)^{8-1} \\ &= (3)^7 \\ &= 2{,}187 \end{aligned}$$

(b) 64, −16, 4, −1, ...

The common ratio is

$$r = \frac{a_2}{a_1} = \frac{-16}{64} = -\frac{1}{4}$$

The general term is

$$a_n = 64\left(-\frac{1}{4}\right)^{n-1}$$

The eighth term of the sequence is

$$\begin{aligned} a_8 &= 64\left(-\frac{1}{4}\right)^{8-1} \\ &= 64\left(-\frac{1}{4}\right)^7 \\ &= -\frac{1}{256} \end{aligned}$$

(c) 16, 24, 36, 54, ...

The common ratio is

$$r = \frac{a_2}{a_1} = \frac{24}{16} = \frac{3}{2}$$

The general term is

$$a_n = 16\left(\frac{3}{2}\right)^{n-1}$$

The eighth term of the sequence is

$$\begin{aligned} a_8 &= 16\left(\frac{3}{2}\right)^{8-1} \\ &= 16\left(\frac{3}{2}\right)^{7} \\ &= \frac{2{,}187}{8} \end{aligned}$$

Geometric Series

A **geometric series** is the sum of the terms in a geometric sequence. If the sequence has a finite number of terms, the formula for the sum is

Formula 3:

$$S_n = \frac{a_1(1-r^n)}{1-r} \qquad (r \neq 1)$$

This form of the formula is used when the number of terms (n), the first term (a_1), and the common ratio (r) are known.

Another formula for the sum of a geometric sequence is

Formula 4:

$$S_n = \frac{a_1 - a_n r}{1-r} \qquad (r \neq 1)$$

This form requires the first term (a_1), the last term (a_n), and the common ratio (r), but does not require the number of terms (n).

Example 1: Find the sum of the first five terms of the geometric sequence in which $a_1 = 3$ and $r = -2$.

First, find the values of a_1, r, and n.

$$a_1 = 3, r = -2, n = 5$$

Use Formula 3:

$$S_n = \frac{a_1(1-r^n)}{1-r}$$

$$S_5 = \frac{3[1-(-2)^5]}{1-(-2)}$$

$$= \frac{3[1-(-32)]}{3}$$

$$= \frac{99}{3}$$

$$= 33$$

Example 2: Find the sum of the geometric sequence for which $a_1 = 48$, $a_n = 3$, and $r = -\frac{1}{2}$.

Use Formula 4:

$$S_n = \frac{a_1 - a_n r}{1-r}$$

$$= \frac{48-3\left(-\frac{1}{2}\right)}{1-\left(-\frac{1}{2}\right)}$$

$$= \frac{48+\frac{3}{2}}{\frac{3}{2}}$$

$$= \frac{\frac{99}{2}}{\frac{3}{2}}$$

$$= 33$$

Example 3: Find a_1 in each geometric series.

(a) $S_n = 244$, $r = -3$, $n = 5$

Use Formula 3:

$$S_n = \frac{a_1(1-r^n)}{1-r}$$

$$244 = \frac{a_1[1-(-3)^5]}{1-(-3)}$$

$$244 = \frac{a_1(244)}{4}$$

$$a_1 = 4$$

(b) $S_n = 15.75$, $r = 0.5$, $a_n = 0.25$

Use Formula 4:

$$S_n = \frac{a_1 - a_n r}{1-r}$$

$$15.75 = \frac{a_1 - (0.25)(0.5)}{1-0.5}$$

$$15.75 = \frac{a_1 - (0.125)}{0.5}$$

$$7.875 = a_1 - 0.125$$

$$a_1 = 8$$

If a geometric series is infinite and $-1 < r < 1$, the formula for its sum becomes

Formula 5:

$$S_n = \frac{a_1}{1-r}$$

If $r > 1$ or if $r < -1$, the infinite series does not have a sum.

Example 4: Find the sum of each of the following geometric series.

(a) 25 + 20 + 16 + 12.8 + ...

First find r.

$$r = \frac{20}{25} = \frac{4}{5}$$

Since $-1 < \frac{4}{5} < 1$, this infinite geometric series has a sum.

Use Formula 5:

$$\begin{aligned} S_n &= \frac{a_1}{1-r} \\ &= \frac{25}{1-\frac{4}{5}} \\ &= \frac{25}{\frac{1}{5}} \\ &= 125 \end{aligned}$$

(b) 3 − 9 + 27 − 81 + ...

First find r.

$$r = \frac{-9}{3} = -3$$

Since $r < -1$, this geometric series does not have a sum.

Summation Notation

A method for writing the sum of a finite number of terms in a sequence is **summation notation.** This notation uses the Greek letter sigma, Σ. When using the sigma notation, the variable defined below the Σ is called the **index of summation.** The lower number is the lower limit of the index (the term where the summation starts), and the upper number is the upper limit of the summation (the term where the summation ends). Consider the following summation notation:

$$\sum_{k=2}^{7} (2k+3)$$

This is read as "the summation of $(2k + 3)$ as k goes from 2 to 7." The replacements for the index are always consecutive integers.

$$\begin{aligned}\sum_{k=2}^{7}(2k+3) &= \overset{k=2}{[2(2)+3]}+\overset{k=3}{[2(3)+3]}+\overset{k=4}{[2(4)+3]}+\overset{k=5}{[2(5)+3]}+\overset{k=6}{[2(6)+3]}+\overset{k=7}{[2(7)+3]}\\ &= 7+9+11+13+15+17\\ &= 72\end{aligned}$$

Example 1: Write out the terms of the following sums; then compute the sum.

(a) $\sum_{j=0}^{5} 3j = \overset{j=0}{3(0)}+\overset{j=1}{3(1)}+\overset{j=2}{3(2)}+\overset{j=3}{3(3)}+\overset{j=4}{3(4)}+\overset{j=5}{3(5)} = 45$

(b) $\sum_{k=0}^{4} 2^k = \overset{k=0}{2^0}+\overset{k=1}{2^1}+\overset{k=2}{2^2}+\overset{k=3}{2^3}+\overset{k=4}{2^4} = 31$

(c) $$\begin{aligned}\sum_{n=1}^{5}\frac{3n+2}{n+1} &= \overset{n=1}{\frac{5}{2}}+\overset{n=2}{\frac{8}{3}}+\overset{n=3}{\frac{11}{4}}+\overset{n=4}{\frac{14}{5}}+\overset{n=5}{\frac{17}{6}}\\ &= \frac{150+160+165+168+170}{60}\\ &= \frac{813}{60}\\ &= \frac{271}{20}\end{aligned}$$

Example 2: Use sigma notation to express each series.

(a) $8 + 11 + 14 + 17 + 20$

This is an arithmetic series with five terms; the first term is 8 and the common difference is 3. Therefore, $a_1 = 8$ and $d = 3$. The nth term of the corresponding sequence is

$$\begin{aligned}a_n &= a_1+(n-1)d\\ &= 8+(n-1)3\\ &= 3n+5\end{aligned}$$

Since there are five terms, the given series can be written as

$$\sum_{n=1}^{5} a_n = \sum_{n=1}^{5}(3n+5)$$

(b) $\frac{2}{3} - 1 + \frac{3}{2} - \frac{9}{4} + \frac{27}{8} - \frac{81}{16}$

This is a geometric series with six terms; the first term is $\frac{2}{3}$ and the common ratio is $-\frac{3}{2}$. Therefore, $a_1 = \frac{2}{3}$ and $r = -\frac{3}{2}$. The nth term of the corresponding sequence is

$$\begin{aligned} a_n &= a_1 r^{n-1} \\ &= \frac{2}{3}\left(-\frac{3}{2}\right)^{n-1} \end{aligned}$$

Since there are six terms in the given series, the sum can be written as

$$\sum_{n=1}^{6} a_n = \sum_{n=1}^{6} \frac{2}{3}\left(-\frac{3}{2}\right)^{n-1}$$

Applications of Sequences and Series

Knowledge of sequences and series is often useful in dealing with everyday familiar situations.

Example 1: If you save 10 cents on the first day of the year and 2 cents more every day for 365 days, how much will you save on the 365th day, and how much will you have saved for the entire year?

This is an arithmetic sequence: 10, 12, 14, 16, ... for 365 days.

$$\begin{aligned} a_n &= a_1 + (n-1)d \\ a_{365} &= 10 + (365-1)(2) \\ a_{365} &= 738 \end{aligned}$$

$$\begin{aligned} S_n &= \frac{n}{2}(a_1 + a_n) \\ S_{365} &= \frac{365}{2}(10 + 738) \\ S_{365} &= 136{,}510 \end{aligned}$$

Note that the solution above is in cents, not dollars.

On the 365th day, $7.38 will be saved, bringing the total sum on that day to $1,365.10.

Chapter Check-Out

Questions

1. Find the twentieth term of each arithmetic sequence:

   (a) 0, 6, 12, ...

   (b) $-\frac{3}{4}, -\frac{2}{4}, -\frac{1}{4}, \ldots$

   (c) 5, 4.4, 3.8, ...

2. Find the sum of multiples of 5 between 4 and 112.
3. Find the eighth term of the following geometric sequences:

   (a) 5, −15, 45, ...

   (b) 4, 20, 100, ...

4. Write out the terms of the sums. Then compute the sums.

   (a) $\sum_{n=2}^{9}(3n-5)$

   (b) $\sum_{w=0}^{6} 3^{w}$

5. Are the following sequences arithmetic, geometric, or neither? Justify your answers.

   (a) 1, 4, 9, 16, 25, 36, ...

   (b) 8, −4, 2, −1, ...

   (c) $\frac{1}{4}, \frac{1}{2}, \frac{3}{4}, \ldots$

6. Explain the difference between a sequence and a series.
7. A classic car is purchased for $50,000. Its value increases 10% yearly. How much will it be worth at the end of 6 years? (***Hint:*** The original amount is the first year.)

Answers

1. **(a)** 114; **(b)** 4; **(c)** -6.4
2. 1,265
3. **(a)** $-10{,}935$; **(b)** 312,500
4. **(a)** 1, 4, 7, 10, 13, 16, 19, 22; 92

   (b) 1, 3, 9, 27, 81, 243, 729; 1,093
5. **(a)** neither; there is no common difference and no common ratio

   (b) geometric; common ratio of $-\frac{1}{2}$

   (c) arithmetic; common difference of $\frac{1}{4}$
6. Answers may vary. A sequence is an ordered list of numbers. A series is the sum of the terms in a sequence.
7. \$80,525.50 (***Note:*** This is a geometric sequence whose $r = 1.1$.)

Chapter 16

TRIGONOMETRY AND TRIGONOMETRIC FUNCTIONS

Chapter Check-In

- ❑ Defining trigonometry
- ❑ Defining and evaluating sine, cosine, and tangent—the principle ratios
- ❑ Understanding radian measures
- ❑ Examining angles of rotation, coterminal angles, and reference angles
- ❑ Defining and evaluating secant, cosecant, and cotangent—the reciprocal ratios
- ❑ Applying trigonometry

Common Core Standard: Trigonometric Functions

Extend the domain of trigonometric functions using the unit circle. Prove and apply trigonometric identities (F.TF).

Common Core Standard: Similarity, Right Triangles, and Trigonometry

Apply trigonometry to general triangles (G.SRT).

Defining Trigonometry

The word **trigonometry** is of Greek origin and translates as "triangle measurement." Trigonometry is the branch of mathematics that is used to find distances, angles, and indirect measurement. The study of trigonometry is defined in terms of right triangles. It is the study of the relationship between the angles and the sides of a triangle, and is often applied to measurement problems in the fields of science, engineering, aerospace, and architecture.

Defining and Evaluating the Principle Trigonometry Ratios

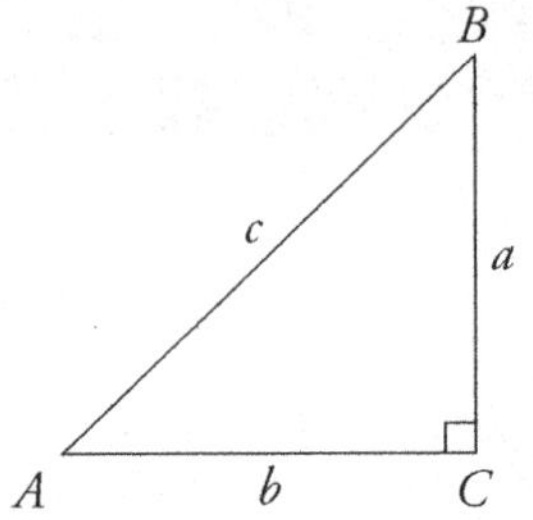

ΔABC is a standard right triangle.

Trigonometric Functions	
sine (sin)	In any right triangle, the sine (sin) of an acute angle is the ratio of the opposite side to the hypotenuse.
cosine (cos)	The cosine (cos) of an acute angle is the ratio of the adjacent side to the hypotenuse.
tangent (tan)	The tangent (tan) of an acute angle is the ratio of the opposite side to the adjacent side.

Use the acronym SOHCAHTOA (sine, cosine, and tangent) to remember the three ratios.

sine = opposite/**h**ypotenuse (soh...)

cosine = adjacent/**h**ypotenuse (...cah...)

tangent = opposite/adjacent (...toa)

$$\sin A = \frac{a}{c} \qquad \sin B = \frac{b}{c}$$

$$\cos A = \frac{b}{c} \qquad \cos B = \frac{a}{c}$$

$$\tan A = \frac{a}{b} \qquad \tan B = \frac{b}{a}$$

Example 1: Express the sin, cos, and tan of an 18° angle in terms of the triangle shown.

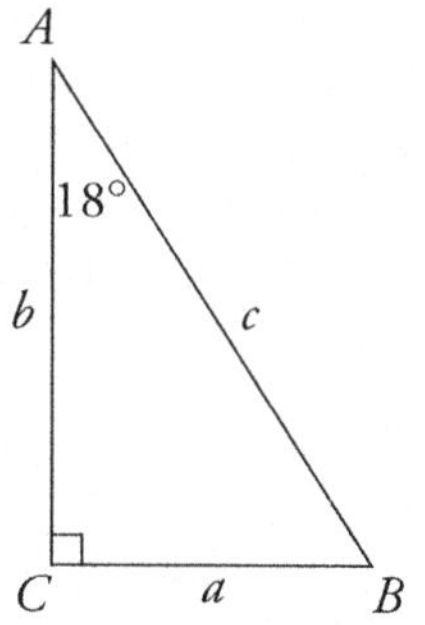

$$\sin 18° = \frac{a}{c},\ \cos 18° = \frac{b}{c},\ \tan 18° = \frac{a}{b}$$

Example 2: Express the sin, cos, and tan of $\angle A$ and $\angle B$ in the triangle below as a decimal.

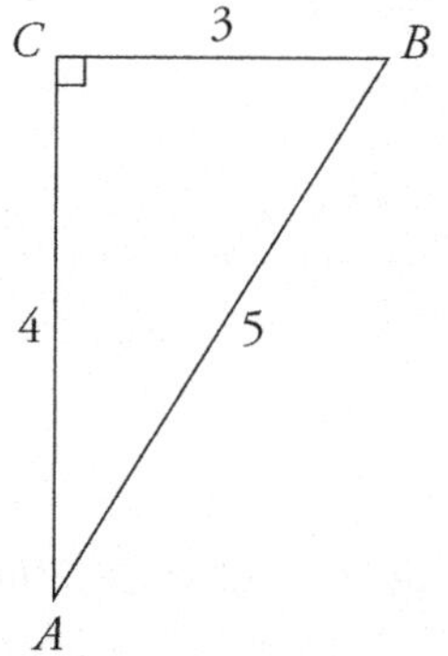

$$\sin A = \frac{3}{5} = 0.6 \qquad \sin B = \frac{4}{5} = 0.8$$

$$\cos A = \frac{4}{5} = 0.8 \qquad \cos B = \frac{3}{5} = 0.6$$

$$\tan A = \frac{3}{4} = 0.75 \qquad \tan B = \frac{4}{3} = 1.\overline{3}$$

Understanding Radian Measures

Angle measures can be expressed in degrees or in radians (rads). Radians are different from degrees but are alternately used as angle measures. In trigonometry, radians are used to describe units of measure for angles instead of degrees. A **radian** is the ratio between the length of an arc to the length of its radius. There are 360 degrees in a full circle and there are 2π radians in the full revolution of a circle.

The conversion between radians and degrees is as follows:

$$\text{degrees} \rightarrow \text{rads, multiply by } \frac{\pi}{180}$$

$$\text{rads} \rightarrow \text{degrees, multiply by } \frac{180}{\pi}$$

Example 1: Convert the following degree measures to radian measures:

(a) 30°

$$30 \cdot \frac{\pi}{180} = \frac{\pi}{6} \text{ rads}$$

(b) 65°

$$65 \cdot \frac{\pi}{180} = \frac{13\pi}{36} \text{ rads}$$

(c) 270°

$$270 \cdot \frac{\pi}{180} = \frac{3\pi}{2} \text{ rads}$$

Example 2: Convert the following radian measures to degree measures:

(a) $\frac{5\pi}{6}$

$$\frac{5\pi}{6} \cdot \frac{180}{\pi} = 150°$$

(b) $\frac{\pi}{4}$

$$\frac{\pi}{4} \cdot \frac{180}{\pi} = 45°$$

(c) 2π

$$2\pi \cdot \frac{180}{\pi} = 360°$$

Angles of Rotation, Coterminal Angles, and Reference Angles

An **angle** is defined as two rays sharing a common endpoint.

A circle centered at the origin whose radius is 1 is called the **unit circle.**

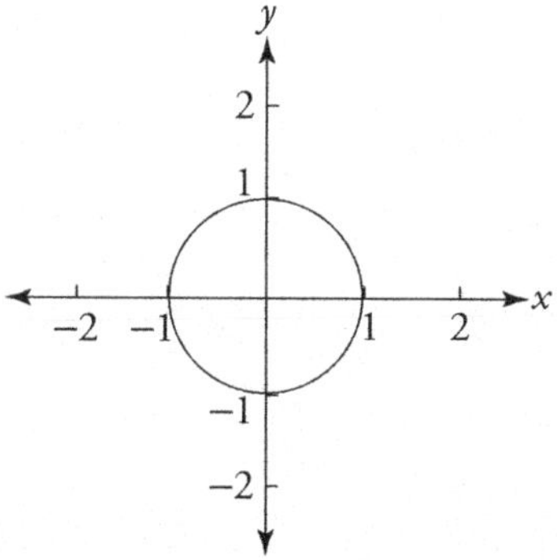

An angle whose vertex is at the origin, or at the center of the unit circle, has a fixed ray on the *x*-axis. This ray is called the *initial side* of the angle. The other ray is called the *terminal side* of the angle.

An angle is formed when both rays begin on the positive *x*-axis and the terminal is rotated. One full rotation represents a 360° angle. Angles that differ by one or more complete rotations are known as **coterminal angles.** See the figure below for a 110° angle and a 470° angle.

A 110° angle is coterminal with a 470° angle.

A 25° angle is coterminal with a 745° angle.

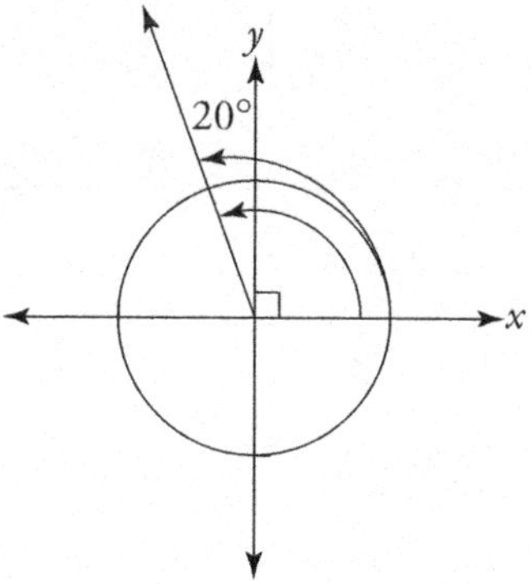

The terminal side of an angle can rotate in the counterclockwise or the clockwise direction. A negative angle implies that the terminal side of the angle rotates clockwise.

A $-45°$ angle is coterminal with a $315°$ angle.

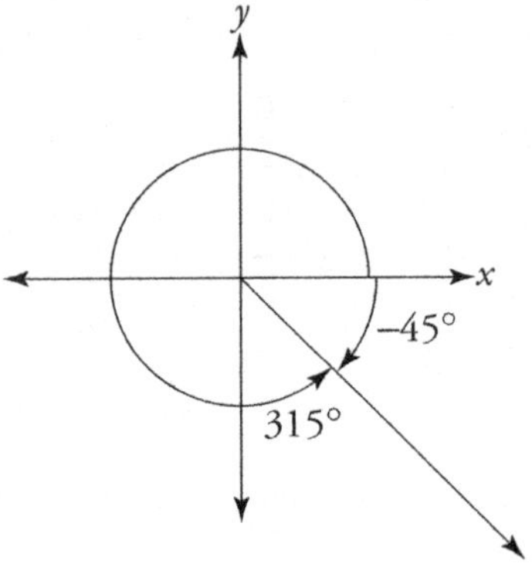

A **reference angle** is the term used to describe the acute angle formed by the terminal side of the angle and the x-axis.

Example 1: $\theta = 225°$

(a) Find its reference angle.

(b) Find one positive and one negative coterminal angle.

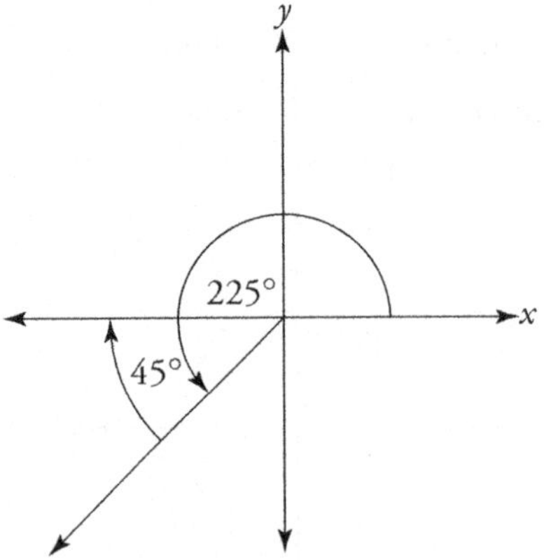

reference angle $= 45°$
a positive coterminal angle $= 585°$
a negative coterminal angle $= -135°$

Defining and Evaluating Reciprocal Ratios: Cosecant, Secant, and Cotangent

Ratios are fractions. Just like fractions have reciprocals, trigonometry ratios have reciprocals. The **cosecant, secant,** and **cotangent** are trigonometric functions that are the reciprocals of the **sine, cosine,** and **tangent**, respectively.

$$\text{cosecant (csc)} = \frac{1}{\text{sine}} \qquad \text{secant (sec)} = \frac{1}{\text{cosine}} \qquad \text{cotangent (cot)} = \frac{1}{\text{tangent}}$$

Example 1: Find all six trigonometry ratios at the angle whose terminal side contains the point $(-3, 5)$.

Sketch a figure and drop a perpendicular from the point to the x-axis.

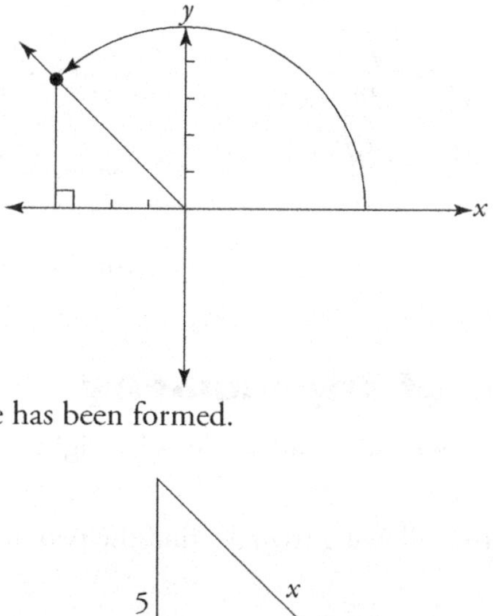

A right triangle has been formed.

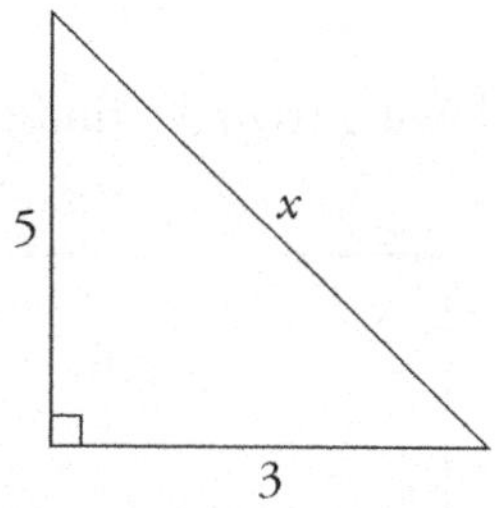

Using the Pythagorean theorem, the hypotenuse is

$$x = \sqrt{3^2 + 5^2}$$
$$x = \sqrt{34}$$

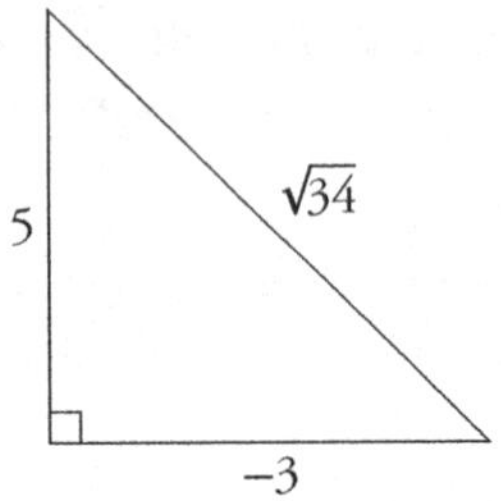

Note: The leg (or the side) of the right triangle cannot be a negative number, but this figure denotes its position on the coordinate plane.

The six trigonometry ratios for the angle whose terminal side contains the point $(-3, 5)$ are

$$\sin A = \frac{5}{\sqrt{34}} = \frac{5\sqrt{34}}{34} \qquad \csc A = \frac{\sqrt{34}}{5}$$

$$\cos A = \frac{-3}{\sqrt{34}} = \frac{-3\sqrt{34}}{34} \qquad \sec A = -\frac{\sqrt{34}}{3}$$

$$\tan A = -\frac{5}{3} \qquad \cot A = -\frac{3}{5}$$

Applications of Trigonometry

When finding unknown sides and angles of a right triangle, we use trigonometry.

Example 1: Given the following triangle, find the two unknowns.

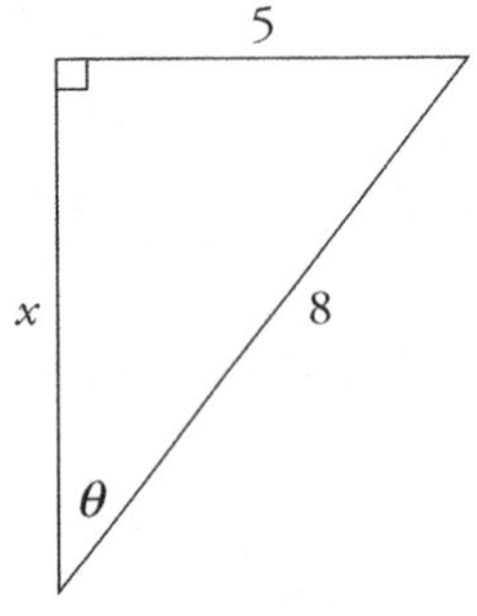

$$\sin \theta = \frac{5}{8} = 0.625$$

Use a calculator to find the angle whose sine = 0.625; $\theta = 38.7°$.

To find x, the Pythagorean theorem or trigonometry could be used.

$$\cos \theta = \frac{x}{8}$$

$$\cos 38.7° = \frac{x}{8}$$

$$0.78 \approx \frac{x}{8}$$

$$6.24 \approx x$$

Example 2: A ladder leans against a building at a 63° angle. The ladder is 20 feet. How high up the building does the top of the ladder touch? Round your answer to the nearest tenth.

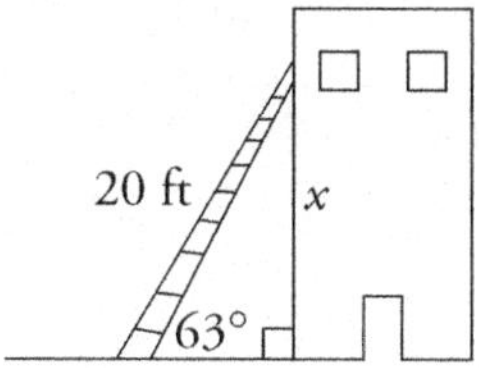

$$\sin 63° = \frac{x}{20}$$

$$0.89 \approx \frac{x}{20}$$

$$17.8 \approx x$$

The top of the ladder touches the building approximately 17.8 feet up.

Example 3: Find the angle of elevation of the sun in the sky when a 30-foot flagpole casts a 40-foot shadow on the ground. Round your answer to the nearest tenth.

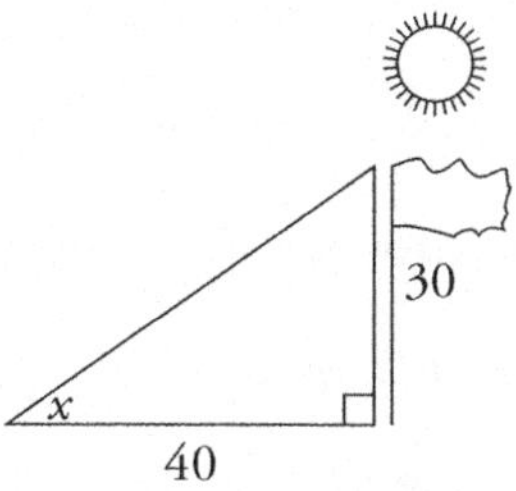

$$\tan x = \frac{30}{40}$$
$$\tan x = 0.75$$
$$x \approx 36.9°$$

The angle of elevation of the sun is approximately 36.9°.

Chapter Check-Out

Questions

1. Based on the following figure, solve for x and y to the nearest hundredth.

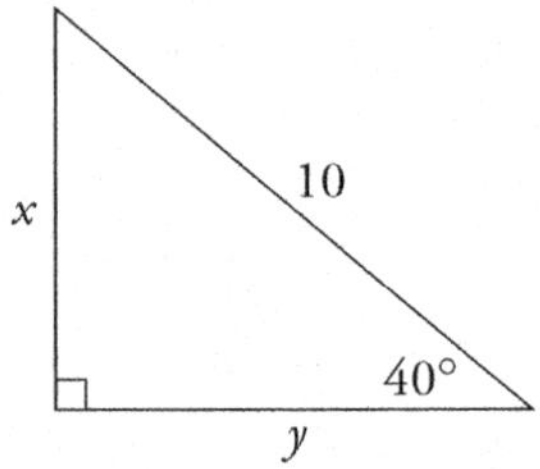

2. Convert $\frac{3\pi}{2}$ to degrees.
3. Convert 140° to radians.
4. Based on the following figure, write all six trigonometry ratios for $\angle B$.

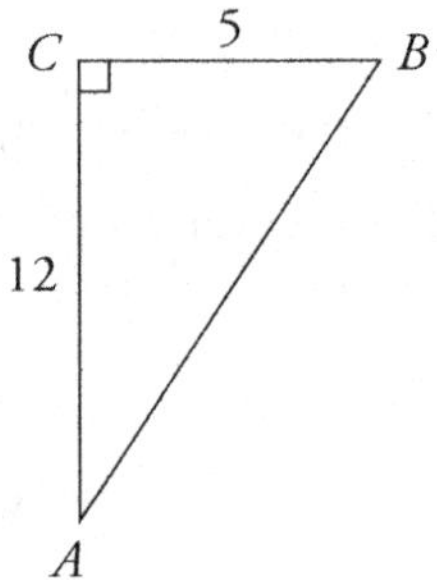

5. Find the reference angle for a 510° angle.

Answers

1. $x \approx 6.43$; $y \approx 7.66$
2. 270°
3. $\frac{7\pi}{9}$
4. $AB = 13$ by the Pythagorean theorem.

$$\sin B = \frac{12}{13} \qquad \csc B = \frac{13}{12}$$

$$\cos B = \frac{5}{13} \qquad \sec B = \frac{13}{5}$$

$$\tan B = \frac{12}{5} \qquad \cot B = \frac{5}{12}$$

5. 30°

Chapter 17
ADDITIONAL TOPICS

Chapter Check-In

- ❑ Defining factorials
- ❑ Using the binomial theorem and coefficients
- ❑ Applying permutations
- ❑ Distinguishing combinations from permutations
- ❑ Calculating the number of possibilities that an event will occur

Common Core Standard: Statistics and Probability

Understand independent and conditional probability and use probability to interpret data. Represent data on two quantitative variables on a scatter plot, and describe how the variables are related. Interpret linear models (S.ID.6.8). Make inferences and justify conclusions (S.IC).

This chapter examines several additional topics in Algebra II. The first to be examined is factorials. Factorials are used in the computing of permutations and combinations, both of which play a part in probability and are examined in this chapter.

Additionally, the topics of binomial theorem and the patterns formed by exponential expansion of binomials—the most notable being Pascal's triangle—are explored in this chapter.

Factorials

A **factorial** is a convenient way of expressing the product of any natural number with all its preceding natural numbers. An exclamation point (!) is the symbol for a factorial. Thus "6!" is read as "six factorial."

$$6! = (6)(5)(4)(3)(2)(1) = 720$$

Since zero is not a natural number, 0! would have no meaning. For the expression 0! to be used in expressing formulas, it is defined as having the value of 1.

$$n! = n(n-1)(n-2)(n-3)\ldots(1)$$

$$0! = 1$$

Example 1: Evaluate $\dfrac{9!}{3!6!}$.

$$\frac{9!}{3!6!} = \frac{(9)(8)(7)(\cancel{6})(\cancel{5})(\cancel{4})(\cancel{3})(\cancel{2})(\cancel{1})}{(3)(2)(1)(\cancel{6})(\cancel{5})(\cancel{4})(\cancel{3})(\cancel{2})(\cancel{1})} = 84$$

Binomial Coefficients and the Binomial Theorem

When a binomial is raised to whole-number powers, the coefficients of the terms in the expansion form a pattern.

$$(a+b)^0 = 1$$

$$(a+b)^1 = 1a + 1b$$

$$(a+b)^2 = 1a^2 + 2ab + 1b^2$$

$$(a+b)^3 = 1a^3 + 3a^2b + 3ab^2 + 1b^3$$

$$(a+b)^4 = 1a^4 + 4a^3b + 6a^2b^2 + 4ab^3 + 1b^4$$

$$(a+b)^5 = 1a^5 + 5a^4b + 10a^3b^2 + 10a^2b^3 + 5ab^4 + 1b^4$$

These expressions exhibit many patterns:

- Each expansion has one more term than the power on the binomial.
- The sum of the exponents in each term in the expansion is the same as the power on the binomial.
- The powers on *a* in the expansion decrease by 1 with each successive term, while the powers on *b* increase by 1.
- The coefficients form a symmetrical pattern.
- Each coefficient entry below the second row is the sum of the closest pair of numbers in the line directly above it.

This triangular array of numbers is called **Pascal's triangle,** named after the French mathematician Blaise Pascal.

Pascal's triangle can be extended to find the coefficients for raising a binomial to any whole-number exponent. This same array could be expressed using the factorial symbol, as shown in the following figure.

row (0)	1 →	$\frac{0!}{0!}$
row (1)	1 1 →	$\frac{1!}{0!1!}$ $\frac{1!}{1!0!}$
row (2)	1 2 1 →	$\frac{2!}{0!2!}$ $\frac{2!}{1!1!}$ $\frac{2!}{2!0!}$
row (3)	1 3 3 1 →	$\frac{3!}{0!3!}$ $\frac{3!}{1!2!}$ $\frac{3!}{2!1!}$ $\frac{3!}{3!0!}$
row (4)	1 4 6 4 1 →	$\frac{4!}{0!4!}$ $\frac{4!}{1!3!}$ $\frac{4!}{2!2!}$ $\frac{4!}{3!1!}$ $\frac{4!}{4!0!}$
row (5)	1 5 10 10 5 1 →	$\frac{5!}{0!5!}$ $\frac{5!}{1!4!}$ $\frac{5!}{2!3!}$ $\frac{5!}{3!2!}$ $\frac{5!}{4!1!}$ $\frac{5!}{5!0!}$
row (n)	1. . . . . .1 →	$\frac{n!}{0!n!}$ $\frac{n!}{1!(n-1)!}$ . . . . $\frac{n!}{(n-1)!1!}$ $\frac{n!}{n!0!}$

In general,

$$(a+b)^n = \frac{n!}{0!n!}(a^n)b^0 + \frac{n!}{1!(n-1)!}(a^{n-1})b^1 + \frac{n!}{2!(n-2)!}(a^{n-2})b^2 + \frac{n!}{3!(n-3)!}(a^{n-3})b^3 + \ldots + \frac{n!}{n!0!}(a^0b^n)$$

The symbol $\binom{n}{k}$, called the **binomial coefficient,** is defined as follows:

$$\binom{n}{k} = \frac{n!}{k!(n-k)!}$$

Therefore,

$$(a+b)^n = \binom{n}{0}a^nb^0 + \binom{n}{1}a^{n-1}b^1 + \binom{n}{2}a^{n-2}b^2 + \ldots + \binom{n}{n}a^0b^n$$

This could be further condensed using sigma notation.

$$(a+b)^n = \sum_{k=0}^{n}\binom{n}{k}a^{n-k}b^k$$

This formula is known as the **binomial theorem.**

Example 1: Use the binomial theorem to express $(x + y)^7$ in expanded form.

$$
\begin{aligned}
(x+y)^7 &= \sum_{k=0}^{7} \binom{7}{k} x^{7-k} y^k \\
&= \binom{7}{0} x^7 y^0 + \binom{7}{1} x^6 y^1 + \binom{7}{2} x^5 y^2 + \binom{7}{3} x^4 y^3 + \binom{7}{4} x^3 y^4 \\
&\quad + \binom{7}{5} x^2 y^5 + \binom{7}{6} x^1 y^6 + \binom{7}{7} x^0 y^7 \\
&= \frac{7!}{0!7!} x^7 y^0 + \frac{7!}{1!6!} x^6 y^1 + \frac{7!}{2!5!} x^5 y^2 + \frac{7!}{3!4!} x^4 y^3 + \frac{7!}{4!3!} x^3 y^4 \\
&\quad + \frac{7!}{5!2!} x^2 y^5 + \frac{7!}{6!1!} x^1 y^6 + \frac{7!}{7!0!} x^0 y^7 \\
&= x^7 + 7x^6 y + 21x^5 y^2 + 35x^4 y^3 + 35x^3 y^4 + 21x^2 y^5 + 7xy^6 + y^7
\end{aligned}
$$

Notice the following pattern:

- The first term $\binom{7}{0} x^7 y^0$
- The second term $\binom{7}{1} x^6 y^1$
- The third term $\binom{7}{2} x^5 y^2$

In general, the *k*th term of any binomial expansion can be expressed as follows:

$$
k\text{th term} = \binom{n}{k-1} x^{n-(k-1)} y^{k-1}
$$

Example 2: Find the tenth term of the expansion $(x + y)^{13}$.

Since $n = 13$ and $k = 10$,

$$k\text{th term} = \binom{n}{k-1} x^{n-(k-1)} y^{k-1}$$

$$10\text{th term} = \binom{13}{9} x^{13-9} y^{9}$$

$$= \frac{13!}{9!4!} x^{4} y^{9}$$

$$= 715x^{4} y^{9}$$

Permutations and Combinations

Multiplication principle for events

The **multiplication principle for events** states that if one event can occur in p different ways, and another independent event can occur in q different ways, then there are pq ways that both events can occur together. Making choices and ordering are not factors in this calculation.

Example 1: How many different ways can a man coordinate a wardrobe if he has a choice of four different pants and two different sport jackets?

The first event (selecting pants) can occur in four different ways. The second event (selecting a sport jacket) can occur in two ways. Therefore, according to the multiplication principle for events, there are $(4)(2) = 8$ different ways for the man to coordinate a wardrobe.

Example 2: In how many ways can five books be arranged on a shelf?

The first space can be filled with any one of the five books, the second space with any of the remaining four books, the third space with any of the remaining three books, the fourth space with either of the remaining two books, and the fifth space with the last book. Therefore, there are $(5)(4)(3)(2)(1) = 120$ ways to arrange five books on a shelf. This can also be written as 5!.

Often, making choices and ordering do become factors in calculation. This is where a knowledge of permutations and combinations becomes important.

Example 3: In how many ways can three out of eight books be arranged on a shelf?

The first space can be filled in any of eight ways, the second space in any of the remaining seven ways, and the third space in any of the remaining six ways. Therefore, there are (8)(7)(6) = 336 ways that three out of eight books can be arranged on a shelf.

Permutations

The arrangement of objects in a *specific order* is called a **permutation.** The number of ways to arrange eight things taken three at a time is written as $P(8, 3)$ or as ${}_8P_3$. $P(n, r)$ or ${}_nP_r$ is read as "the permutation of n things taken r at a time."

$$ {}_nP_r = P(n,r) = \frac{n!}{(n-r)!} $$

If any of the objects in a permutation are repeats, a different formula is used. The number of permutations of n objects of which p are alike and q are alike is

$$ \frac{n!}{p!q!} $$

Example 4: How many different ways can the letters in the word "Mississippi" be arranged?

"Mississippi" has 11 letters. Since "i" is repeated four times, "s" is repeated four times, and "p" is repeated two times, there are $\frac{11!}{4!4!2!} = 34{,}650$ different ways that the letters in the word "Mississippi" can be arranged.

Combinations

When the order in which objects are chosen is *not* important, the arrangement is called a **combination.** The combination of n things taken r at a time is written as $C(n, r)$ or ${}_nC_r$. A formula for finding the number of combinations of n objects taken r at a time is given by the following:

$$ {}_nC_r = C(n,r) = \frac{n!}{r!(n-r)!} $$

Note that this is the same as the binomial coefficient formula.

Example 5: A class has 16 students. How many groups consisting of four students can be formed?

This is a combinations problem, since the order in which the students in a group are chosen is not important. In this problem, $n = 16$ and $r = 4$.

$$\begin{aligned} C(16,4) &= \frac{16!}{4!(16-4)!} \\ &= \frac{(16)(15)(14)(13)\cancel{(12!)}}{(4)(3)(2)(1)\cancel{(12!)}} \\ &= 1,820 \end{aligned}$$

There are 1,820 different ways to form the groups.

Example 6: A committee in Congress consists of nine Republicans and six Democrats. In how many ways can a subcommittee be chosen if it must contain four Republicans and three Democrats?

There are $C(9, 4)$ ways of choosing the four Republicans out of the nine Republicans. There are $C(6, 3)$ ways of choosing the three Democrats out of the six Democrats. By the multiplication principle of events, there are $C(9, 4) \cdot C(6, 3)$ ways of choosing the subcommittee.

$$\begin{aligned} C(9,4)\cdot C(6,3) &= \frac{9!}{4!(9-4)!}\cdot\frac{6!}{3!(6-3)!} \\ &= \frac{(9)(8)(7)(6)\cancel{(5!)}}{(4)(3)(2)(1)\cancel{(5!)}}\cdot\frac{(6)(5)(4)\cancel{(3!)}}{(3)(2)(1)\cancel{(3!)}} \\ &= (126)(20) \\ &= 2,520 \end{aligned}$$

There are 2,520 different possible subcommittees.

Calculating the Number of Possibilities

Decision making is often made easier when the number of outcomes or choices is known. Calculating the number of possibilities that an event will occur is directly related to making good decisions.

Example 1: Imagine that you are in a Las Vegas casino playing five-card draw poker and making a bet on drawing (at most) one king in a five-card hand. The casino is using a standard 52-card deck. How many possible five-card hands containing at most one king are there?

To have exactly one king, you would have to have one of the four kings, and 4 of the remaining 48 cards that are not kings. This calculation becomes $\frac{4!}{3!1!} \cdot \frac{48!}{44!4!} = 778,320.$

This is based on the number of combinations of four kings taken 1 at a time, multiplied by the number of combinations of 48 cards taken 4 at a time.

To hold no king in your hand, the calculation would be $\frac{48!}{43!5!} = 1,712,304$, based on the number of combinations of 48 cards taken 5 at a time.

The total amount of possible five-card hands containing at most one king is the sum of those two: 778,320 + 1,712,304 = 2,490,624.

Now knowing there are that many possible combinations, would you bet money that you will be dealt at most one king in a five-card hand?

To make your decision on the bet, calculate the percentage of getting the number of ways to have one king divided by the total number of hands. The total number of hands is ${}_{52}C_5 = \frac{52!}{5!47!}$ or 2,598,960.

$$\frac{\text{number of ways to have one king}}{\text{total number of hands possible}} = \frac{2,490,624}{2,598,960} \approx 0.958$$

You'd have about a 96% chance, so yes, make that bet!

Chapter Check-Out

Questions

1. Evaluate $\frac{7!}{6!}$.
2. In how many ways can seven bicycles be arranged in a bike rack with seven slots?
3. In how many ways can four of seven cars be parked in a driveway that has four parking spaces?
4. A bridge club has 20 members. How many different foursomes may be formed?
5. You are going to plant a vegetable garden. You have 18 different kinds of vegetable seeds, but you are going to plant exactly three different kinds of vegetables. How many different combinations of vegetables will you be choosing from?

Answers

1. 7
2. $5{,}040 = 7!$
3. $840 = \frac{7!}{3!} = {}_7P_4$
4. $4{,}845 = {}_{20}C_4$
5. $816 = {}_{18}C_3$

Chapter 18

APPLICATION PROBLEMS

Chapter Check-In

- ❑ Learning strategies for word problem solutions
- ❑ Solving simple interest and compound interest problems
- ❑ Solving mixture problems
- ❑ Solving motion problems
- ❑ Solving work problems
- ❑ Solving arithmetic and geometric series problems
- ❑ Building and interpreting functions

Common Core Standard: Seeing Structure in Expressions

Interpret expressions that represent a quantity in terms of its context. Use the structure of an expression to identify ways to rewrite it. Write expressions in equivalent forms to solve problems (A.SSE). Understand solving equations as a process of reasoning and explain the reasoning (A.REI). Interpret functions that arise in applications in terms of context (F.IF).

If there is any answer to the question, "What good will Algebra II do me in the real world?" this chapter has it. In many occupations, from carpentry to pharmacology, from nursing to aviation, and many other fields, word problems may be a part of everyday life or may be encountered on occasion. Whichever the case, your understanding of how to deal with word problems will prove to be invaluable and will validate the need to learn Algebra II.

This chapter explains the basic strategies to use when approaching word problems, and also provides specific tactics and formulas that apply to specific types of word problems.

General Strategies

Although word problems differ from one another, you can use some general strategies to solve them:

1. Read the problem carefully, several times if necessary, to familiarize yourself with what is being asked. Remember to answer "all parts" of the problem.
2. Select a variable(s) to represent the unknown(s) in the problem.
3. Use visual organizers. If possible, draw a diagram to illustrate the facts in a problem or create a chart to organize the given information.
4. Translate the problem into an algebraic equation using the variable(s) chosen. If you're uncertain about how to begin, look for familiar words to help you understand the overall scope and context of the problem.
5. Solve the algebraic equation.
6. Check your result.
7. Reread the problem; see how the answer to the algebraic equation is used to answer the question asked. Check to see whether the answer is reasonable.
8. Finally, answer the question.

Simple Interest

The formula to solve simple interest problems is

$$I = PRT$$

Example 1: Jim has $10,000 to invest. He will invest part at 9% annual interest and the remaining part at 12% annual interest. After 1 year, he expects to earn $165 more from the 9% investment than from the 12% investment. How much will he invest at each rate?

Let x = amount invested at 9%. Then

$$0.09x = \text{amount of annual interest at } 9\%$$

$$10{,}000 - x = \text{amount invested at } 12\%$$

$$\text{Then } 0.12(10{,}000 - x) = \text{amount of annual interest at } 12\%$$

"He expects to earn $165 more from the 9% investment than from the 12% investment" is translated into the following.

Interest at 9% is $165 more than interest at 12%.

$$0.09x = 0.12(10{,}000 - x) + 165$$
$$0.09x = 1{,}200 - 0.12x + 165$$
$$0.21x = 1{,}365$$
$$x = 6{,}500$$

Jim will invest $6,500 at 9% and $3,500 at 12%.

Compound Interest

The formula necessary to solve most compound interest problems is

$$A = P\left(1 + \frac{r}{n}\right)^{nt}$$

where A = future value of the investment

P = principal (original investment)

r = annual interest rate

t = time in years of the investment

n = number of times the investment is compounded annually

Example 1: How long would it take for an investment of $3,500 to become $4,200 if it is invested in an account that earns 6% interest compounded monthly?

$$A = 4{,}200 \qquad P = 3{,}500 \qquad r = 0.06 \qquad n = 12$$

$$A = P\left(1 + \frac{r}{n}\right)^{nt}$$
$$4{,}200 = 3{,}500\left(1 + \frac{0.06}{12}\right)^{12t}$$
$$4{,}200 = 3{,}500(1 + 0.005)^{12t}$$
$$4{,}200 = 3{,}500(1.005)^{12t}$$
$$\frac{4{,}200}{3{,}500} = (1.005)^{12t}$$
$$1.2 = (1.005)^{12t}$$

$$\log(1.2) = \log(1.005)^{12t}$$
$$\log(1.2) = 12t[\log(1.005)]$$
$$\frac{\log(1.2)}{12\log(1.005)} = t$$
$$3.05 \approx t$$

The \$3,500 investment would have become \$4,200 in about 3.05 years, or just over 3 years and 2 weeks.

Mixture

Example 1: A radiator's capacity is 16 gallons. If the radiator is full of a 40% antifreeze solution, how many gallons must be drained and replaced with a 75% antifreeze solution to obtain a full radiator of 45% antifreeze solution?

A diagram is helpful here.

	original radiator mixture		x gallons drained off		x gallons replaced		new radiator mixture
	16 gallons 40% antifreeze	−	x gallons 40% antifreeze	+	x gallons 75% antifreeze	=	16 gallons 45% antifreeze
antifreeze	0.40(16)	−	0.40(x)	+	0.75(x)	=	0.45(16)

$$0.40(16) - 0.40x + 0.75x = 0.45(16)$$
$$6.4 + 0.35x = 7.2$$
$$0.35x = 0.8$$
$$x = \frac{80}{35} = \frac{16}{7} = 2\frac{2}{7}$$
$$\approx 2.29 \text{ gallons}$$

So $\frac{16}{7}$ or $2\frac{2}{7}$ gallons (approximately 2.29 gallons) of the 40% antifreeze solution must be drained and replaced.

Motion

Motion problems are also known as *distance* problems. These types of problems calculate the distance, rate of speed, and time that it takes to travel from one point to another.

The formula for motion problems is

$$d = rt$$

$$d = \text{distance traveled}$$

$$r = \text{rate of travel}$$

$$t = \text{time traveled}$$

Example 1: A train leaves Chicago at 11:00 a.m. traveling east at a speed of 40 mph. Two hours later, a second train leaves Chicago on a parallel track traveling in the same direction as the first train. The second train travels at a rate of 50 mph. Assuming that neither train stops, at what time will the second train catch up to the first train? At that time, how far has each train traveled?

Drawing a diagram helps in understanding the situation. Constructing a chart helps in organizing the data.

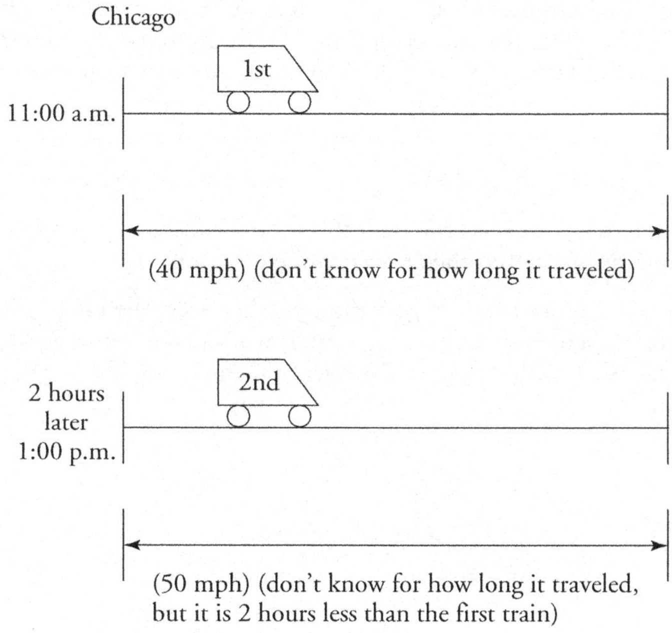

Since both trains began at the same point and will be at the same point when the second train catches up to the first, they will have traveled the same distance. Let t equal the number of hours the first train travels. Since the second train leaves 2 hours later, it will have traveled for 2 hours less. Therefore, $t - 2$ equals the number of hours the second train travels.

	r	$\times$	t	$=$	d
first train	40 mph	$\times$	t hr	$=$	$40t$ mi
second train	50 mph	$\times$	$(t-2)$ hr	$=$	$50(t-2)$ mi

The trains travel the same distance, so

$$50(t-2) = 40t$$
$$50t - 100 = 40t$$
$$10t = 100$$
$$t = 10$$

Therefore, it will be 10 hours after the first train leaves before the second train catches up to the first train. *But this is not the question!* The question asks, "At what time will the second train catch up to the first train?" The second train will catch up to the first train 10 hours after 11:00 a.m., which is 9:00 p.m. The second question asks, "At that time, how far has each train traveled?" Replace t into either distance category and evaluate.

$$40t = 40(10) = 400 \text{ or } 50(t - 2) = 50(10 - 2) = 400$$

Therefore, each train has traveled 400 miles. Notice that the answer to the algebra is not always the answer to the question.

Example 2: Susan's boat can go 9 mph in still water. She can go 44 miles downstream in a river in the same time it would take her to go 28 miles upstream. What is the speed of the river? The following figure shows this situation.

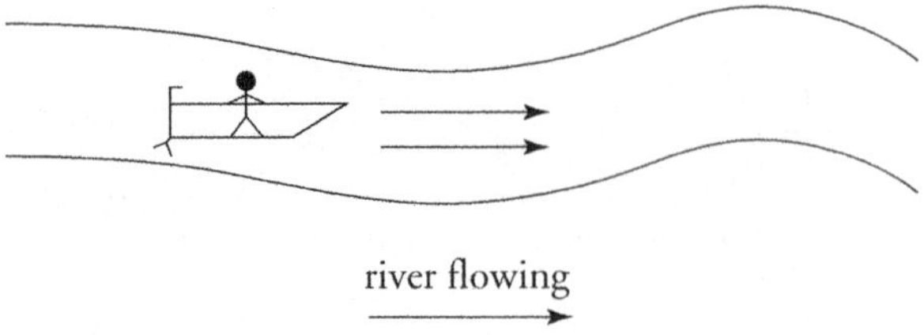

When Susan goes in the same direction as the river, her speed increases by the speed of the river. When she goes against the river, upstream, her speed decreases by the speed of the river.

Let r = the speed of the river.

$$9+r = \text{Susan's speed downstream}$$
$$9-r = \text{Susan's speed upstream}$$

Organize the information in a chart.

	r	$\times$	t	$=$	d
downstream	$(9+r)$ mph	$\times$	t_1	$=$	44 mi
upstream	$(9-r)$ mph	$\times$	t_2	$=$	28 mi

Since $(r)(t) = d$, then $t = \frac{d}{r}$. So $t_1 = \frac{44}{9+r}$, $t_2 = \frac{28}{9-r}$.

In the problem, the key phrase is "in the same time." So set the times t_1 and t_2 equal.

$$\frac{44}{9+r} = \frac{28}{9-r}$$
$$44(9-r) = 28(9+r)$$
$$11(9-r) = 7(9+r)$$
$$99-11r = 63+7r$$
$$-18r = -36$$
$$r = 2$$

Therefore, the river's speed is 2 mph.

Work

The formula for work problems is

$$w = rt$$

w = amount of work done

r = rate of work

t = time spent performing job

For example, if a person can do a job in 4 hours, then his or her rate of work is

$$\frac{1 \text{ job}}{4 \text{ hours}}$$

A person's rate of work is the reciprocal of how long it takes to accomplish the job.

Example 1: George can mow a lawn in 3 hours. Joyce can do it in 2 hours. If they work together, how long will it take?

George needs 3 hours to do the job.

$$\text{George's rate of work} = \frac{1 \text{ job}}{3 \text{ hours}}$$

Joyce needs 2 hours to do the job.

$$\text{Joyce's rate of work} = \frac{1 \text{ job}}{2 \text{ hours}}$$

Let $t =$ the time it takes to mow the lawn together and create a chart.

	r	$\times$	t	$=$	w
George	$\frac{1 \text{ job}}{3 \text{ hr}}$	$\times$	t hr	$=$	$\frac{1}{3}t$ job
Joyce	$\frac{1 \text{ job}}{2 \text{ hr}}$	$\times$	t hr	$=$	$\frac{1}{2}t$ job

Together, they accomplish one job. Therefore,

$$\frac{1}{3}t + \frac{1}{2}t = 1$$

$$6\left(\frac{1}{3}t + \frac{1}{2}t\right) = 6(1)$$

$$2t + 3t = 6$$

$$5t = 6$$

$$t = \frac{6}{5}$$

Together, they will finish the job in $\frac{6}{5}$ or $1\frac{1}{5}$ hours, which is the same as 1 hour and 12 minutes.

Example 2: It takes Angela 3 hours longer to paint a fence than it takes Gary. When they work together, it takes them 2 hours to paint the fence. How long would it take each of them to paint the fence alone?

Let x = the number of hours it takes Gary to do the job alone.

$$\frac{1 \text{ job}}{x \text{ hours}} = \text{Gary's rate of work}$$

Let $x + 3$ = the number of hours it takes Angela to do the job alone (Angela takes 3 hours longer than Gary does).

$$\frac{1 \text{ job}}{x+3 \text{ hours}} = \text{Angela's rate of work}$$

They complete the job in 2 hours. Create a chart to show this information.

	r	$\times$	t	$=$	w
Gary	$\frac{1}{x}\frac{\text{job}}{\text{hr}}$	$\times$	2 hr	$=$	$2\left(\frac{1}{x}\right)$ job
Angela	$\frac{1}{x+3}\frac{\text{job}}{\text{hr}}$	$\times$	2 hr	$=$	$2\left(\frac{1}{x+3}\right)$ job

Together they finish one job:

$$2\left(\frac{1}{x}\right)+2\left(\frac{1}{x+3}\right)=1$$

$$\frac{2}{x}+\frac{2}{x+3}=1$$

$$\cancel{x}(x+3)\cdot\frac{2}{\cancel{x}}+x\cancel{(x+3)}\cdot\frac{2}{\cancel{x+3}}=x(x+3)(1)$$

$$2x+6+2x=x^2+3x$$

$$0=x^2-x-6$$

$$0=(x-3)(x+2)$$

Therefore, $x = 3$ or $x = -2$.

Since x represents a length of time, $x = -2$ has no meaning in this problem. It takes Gary 3 hours and Angela $3 + 3 = 6$ hours to paint the fence alone.

Arithmetic/Geometric Series

Example 1: A grocery store display of soup cans has 16 rows with each row having one less can than the row below it. If the bottom row has 28 cans, how many cans are in the display?

This is the sum of an arithmetic sequence with 16 terms. The sixteenth term is 28, and the common difference is 1. In order to use the formula $S_n = \frac{n}{2}(a_1 + a_n)$, a_1 must be found first.

$$a_n = a_1 + (n-1)d, \text{ where } a_n = 28,\ n = 16, \text{ and } d = 1$$

$$28 = a_1 + (16-1)(1)$$

$$28 = a_1 + 15$$

$$13 = a_1$$

Now, plug the values into the formula and solve.

$$S_{16} = \frac{16}{2}(13+28) = 328$$

There are 328 cans in the display.

Another approach could be to think of this as the sum of an arithmetic sequence with 16 terms, the first of which is $a_1 = 16$, and whose common difference is $d = -1$. Then the direct use of the equation $S_n = \frac{n}{2}[2a_1 + (n-1)d]$ yields the answer.

$$\begin{aligned} S_{16} &= \frac{16}{2}[2(28)+(16-1)(-1)] \\ &= 8(56-15) \\ &= 8(41) \\ &= 328 \end{aligned}$$

Example 2: A ball is dropped from a table that is 24 inches high. The ball always rebounds three-fourths of the distance fallen. Approximately how far, in feet, will the ball have traveled when it finally comes to rest?

The following figure shows this situation.

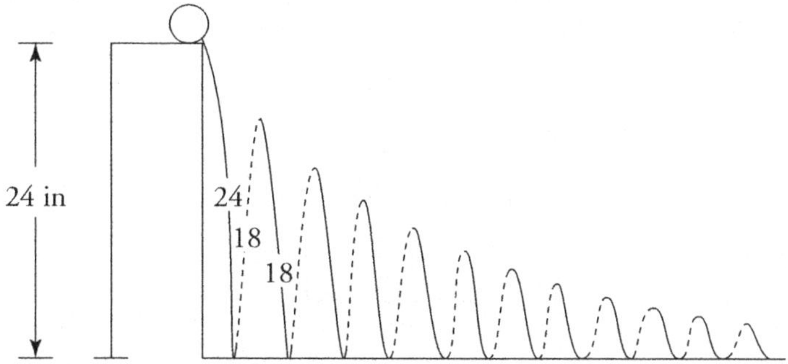

Notice that this problem actually involves two infinite geometric series. One series involves the ball falling, while the other series involves the ball rebounding.

Falling, $a_1 = 24,\ r = \frac{3}{4}$

Rebounding, $a_1 = 24\left(\frac{3}{4}\right) = 18,\ r = \frac{3}{4}$

Use the formula for an infinite geometric series with $-1 < r < 1$.

$$S = \frac{a_1}{1-r}$$

$$S_{\text{falling}} = \frac{24}{1-\frac{3}{4}} = \frac{24}{\frac{1}{4}} = 96$$

$$S_{\text{rebounding}} = \frac{18}{1-\frac{3}{4}} = \frac{18}{\frac{1}{4}} = 72$$

$$S_{\text{falling}} + S_{\text{rebounding}} = 168$$

The ball will travel approximately 168 inches before it finally comes to rest. Converting to feet, the ball will travel $168 \div 12 = 14$ feet.

Building and Interpreting Functions

Translating a word problem is the same concept as building a function from given information. The function can be exponential, linear, or quadratic.

Example 1: The perimeter of a rectangle is 36 feet. Express the area of this rectangle as a function of its width. Find the dimensions of the rectangle that produces the greatest area.

$$A = lw$$

$$f(x) = x(18-x)$$

$$f(x) = -x^2 + 18x$$

$$f(x) = -(x^2 - 18x + 81) - 81$$

$$f(x) = -(x-9)^2 - 81$$

This function is a quadratic in vertex form. This is the function that expresses the area of the rectangle in terms of its width, x, whose value is 9 when the area is greatest at 81 square feet.

Example 2: The function that expresses the degrees in Celsius, given the degrees in Fahrenheit is $f(x) = \frac{5}{9}(x-32)$, where x is the degrees in Fahrenheit.

(a) Given a Fahrenheit temperature of 68 degrees, find the equivalent degrees in Celsius.

$$f(68) = \frac{5}{9}(68-32)$$

$$f(68) = \frac{5}{9}(36)$$

$$f(68) = 20$$

A Fahrenheit temperature of 68 degrees is equivalent to a 20 degree Celsius temperature.

(b) Interpret the ordered pair (32, 0) as a point on the graph of the function $f(x) = \frac{5}{9}(x-32)$.

The point (32, 0) is interpreted as a zero degree Celsius temperature being the equivalent of a Fahrenheit temperature of 32 degrees. This function is a linear function.

Chapter Check-Out

Questions

1. Sandy has $2,000 to invest. She invests part of it at 9% annual interest and the rest at 13%. At the end of a year, she has a total of $2,232. How much did she invest at each rate?
2. Bill wants to mix $5.20/lb cashews with $7.50/lb pecans to make 12 pounds of a mixture that will sell for $6.00/lb. How many pounds of cashews will be in the mixture?
3. A plane leaves City A at 9:00 a.m., heading due east at 530 mph toward City B, which is 2,000 miles away. At the same exact time, a plane leaves City B, heading due west at 450 mph toward City A. At what time (to the nearest minute) will the two planes pass each other?
4. Bill can paint Mrs. Jones's fence in 4 hours. Sue can do it in 3 hours. How long would it take Bill and Sue working together to complete the job?
5. A ball is dropped straight down and rebounds one-half of its preceding height on each bounce. If the ball started out at a height of 128 feet, how high does the ball bounce on its ninth rebound?

Answers

1. $700 at 9%, $1,300 at 13%
2. 7.83 pounds
3. 11:02 a.m.
4. $1\frac{5}{7}$ hours or 1 hour 43 minutes
5. $\frac{1}{4}$ foot

REVIEW QUESTIONS

Use these review questions to practice what you've learned in this book. After you work through the review questions, you're well on your way to achieving your goal of understanding the basic concepts and strategies of Common Core Mathematics Algebra II.

Questions

Chapter 1

1. Solve for x: $3x - 15 \geq 30$.

   A. $x \leq 5$

   B. $x \geq 5$

   C. $x > 15$

   D. $x \geq 15$

2. A parking lot is shaped like a trapezoid with an area of 152 square feet. One of its bases is 14 feet long, and its height is 8 feet. Find the length of the other base.

Chapter 2

3. A line passes through points (–6, 4) and (6, 16). Which of the following is an equation of that line?

   A. $-6x = 4y + b$

   B. $16y = 4x$

   C. $4x - 6y = 10$

   D. $x + 10 = y$

4. Find the y-intercept of the line containing the points (2, –3) and (1, 2).

5. Write the equation in standard form of the line that is perpendicular to the line $3x + 4y = 12$ and has the same y-intercept as the line whose equation is $2x - 3y = 6$.

6. Which of these lines is steeper: a line whose slope is $\frac{3}{8}$ or a line whose slope is $\frac{1}{2}$? Explain your reasoning.

7. The temperature in a classroom changed at a constant rate. Using x as the time in minutes since class began and y as the temperature, find the rate of change in temperature in degrees per minute. The temperature was 70 degrees 15 minutes after class began, and 100 degrees 25 minutes after class began.

Chapter 3

For questions 8–13, solve the following systems of equations.

8. Solve this system of equations by graphing: $\begin{cases} 2x+4y=-22 \\ 5y-6x=-2 \end{cases}$

   A. $x = 3, y = -4$

   B. $x = 3, y = 4$

   C. $x = -3, y = -4$

   D. $x = -4, y = -3$

9. Solve for x and y by graphing: $\begin{cases} 3x-8=y \\ 3x+6y=15. \end{cases}$

10. Solve for x and y by elimination: $\begin{cases} 4x+3y=3 \\ 4y=3x-21. \end{cases}$

11. Solve for x and y using Cramer's Rule: $\begin{cases} 2x+3y=18 \\ 3x-4y=-75. \end{cases}$

12. Solve for x and y by any method: $\begin{cases} 5x-3y=2 \\ 5y-9x=-6. \end{cases}$

13. Solve for x and y by any method: $\begin{cases} 3x+4y=1 \\ 2y-3x=-13. \end{cases}$

Chapter 4

14. Solve the following system of equations for x, y, and z by using Cramer's Rule.

$$\begin{cases} y+4=5z \\ 7z+15-x=28 \\ 4z+11=x+3y \end{cases}$$

A. $x = 1, y = 6, z = 2$

B. $x = 2, y = 5, z = 7$

C. $x = 3, y = 7, z = 5$

D. $x = 8, y = 4, z = 6$

15. Solve this system for x, y, and z using matrices, elimination, or Cramer's Rule.

$$\begin{cases} 2x+4y=6-3z \\ x-3y-2z=-7 \\ x-2y=-5-z \end{cases}$$

A. $x = 0, y = -2, z = 1$

B. $x = -1, y = 2, z = 0$

C. $x = -1, y = -2, z = 0$

D. $x = 1, y = 0, z = 2$

Chapter 5

16. Find the difference:

$$(6 - 3x^3 + 7x^2 - 5x) - (6x^4 - 3x^3 + 5x - 3)$$

A. $-6x^4 + 7x^2 - 10x + 9$

B. $6x^4 + 7x^2 + 10x - 3$

C. $-6x^4 + 7x^2 + 3$

D. $-6x^4 + 7x^2 + 10x + 9$

17. Divide the following by using synthetic division. Express the remainder (if any) as a rational number.

$$\frac{x^4 - 3x^2 - 5x + 20}{x+5}$$

A. $x^2 - 5x + 22 - \dfrac{115}{x+5}$

B. $x^2 - 5x + 22 - 115$

C. $x^3 - 5x^2 + 22x - 115 + \dfrac{595}{x+5}$

D. $x^4 - 5x^3 + 22x - 115 + \dfrac{595}{x+5}$

Chapter 6

18. Factor completely: $6x^2 - x - 15$.

19. Factor completely: $12y^2 - 28y + 16$.

Chapter 7

20. Simplify: $\dfrac{30x^2 + 55x + 15}{14x^2 - 76x + 30} \div \dfrac{6x^2 + 11x + 3}{21x^2 - 114x + 45}$.

A. 6.5

B. 7

C. 7.5

D. 8

21. According to Hooke's Law, the force needed to stretch a spring is proportional to the amount the spring is stretched. If 50 pounds causes a spring to stretch 5 inches, how many inches will the spring stretch when a force of 120 pounds is applied? (***Hint:*** This is a direct variation problem.)

22. Describe and find the asymptotes for the graph of the function $f(x) = \dfrac{4}{x^2 - x - 6}$.

Chapter 8

23. If $f(x) = x^2 + 7x + 5$, find each the following:

(a) $f(7)$

(b) $f(-3)$

(c) $f(0)$

24. If $f(x) = 3x - 7$, find $f^{-1}(x)$.

Chapter 9

25. Find all the zeros for the function $f(x) = 3x^3 + x^2 - 12x - 4$.

A. $\frac{1}{2}, \frac{1}{3}, 2$

B. $-\frac{1}{3}, -2, 2$

C. $-1, 0, 2$

D. $-\frac{2}{3}, -2, 2$

Chapter 10

26. Simplify: $\left(2-3\sqrt{-98}\right)+\left(4\sqrt{-18}\right)$.

Chapter 11

27. Solve for b: $\sqrt{11b+3}-2b=0$.

28. Solve for x: $x^2 - 9x + 18 = 0$.

29. Solve for x: $4x^2 + 9x + 2 = 0$.

30. Solve for x: $2x^3 + 3x^2 - 8x - 12 = 0$.

31. Explain why even roots of negative numbers do not exist in the real number system.

Chapter 12

32. Find the focus, directrix, vertex, and axis of symmetry for the following parabola: $y^2 = 4x + 4y - 16$.

33. Find the equation of the ellipse with center $(-2, 3)$, whose major axis is 12 units long and is parallel to the y-axis and whose minor axis is 8 units long.

Chapter 13

34. Solve for a and b: $\begin{cases} ab = 2 \\ a^2 + b^2 = 4 \end{cases}$

35. Solve for x and y: $\begin{cases} \frac{1}{x} = \frac{1}{y} + \frac{1}{4} \\ x \cdot y = 8 \end{cases}$

Chapter 14

36. Solve $4^{2x} = 17$ for x and round the answer to the hundredths place.

A. 0.99

B. 1.00

C. 1.01

D. 1.02

Chapter 15

37. Find the fifteenth term of the following arithmetic sequence:

1.00, 1.25, 1.50, 1.75, . . .

A. 4.25

B. 4.50

C. 4.75

D. 5.00

Chapter 16

38. A tent has a center pole that is 6 feet tall and a square floor that is 8 feet wide. Find the measure of the angle the tent side forms with the ground. Round your answer to the nearest tenth.

39. Find the approximate measures of the two acute angles in a right triangle whose legs are 7 and 24 units long. Round your answer to the nearest tenth.

40. A 20-foot guy wire is attached to the ground from the top of a 15-foot telephone pole. How far from the base of the pole is the wire attached, and at what angle? Round your answers to the nearest tenth.

Chapter 17

41. How many different six-digit license tags can be printed if the first digit cannot be zero, and no digit may be used a second time?

42. How many permutations are there of the letters "u, v, w, x, y, z"?

Chapter 18

43. Phil takes 2 hours to paint 500 fence pickets and Frank takes 3 hours to paint 450 fence pickets. How long will they take, working together, to paint 1,000 fence pickets? Round your answer to the nearest tenth.

44. How many fluid ounces of distilled water must be added to 100 fluid ounces of a 15% sugar solution to make a solution that is 10% sugar?

Answers

1. D

2. 24 feet

3. D

4. The y-intercept is 7. It is the point (0, 7).

5. $4x - 3y = 6$

6. The line whose slope is $\frac{1}{2}$ is steeper. This line rises 1 unit for every 2 units that it runs, or rises 4 units for every 8 units that it runs. The other line rises 3 units for every 8 units that it runs.

7. The rate of change is 3 degrees per minute.

8. C

9. (3, 1)

10. (3, –3)

11. (–9, 12)

12. (4, 6)

13. (3, –2)

14. A

15. B

16. A

17. C

18. $(2x + 3)(3x - 5)$

19. $4(y - 1)(3y - 4)$

20. C

21. 12 inches

22. vertical asymptotes at $x = -2$ and $x = 3$; horizontal asymptote at $y = 0$

23. **(a)** 103; **(b)** –7; **(c)** 5

24. $\frac{x+7}{3}$

25. B

26. $2-9i\sqrt{2}$

27. $b = 3$ or $b = -\frac{1}{4}$

28. $x = 3$ or $x = 6$

29. $x = -\frac{1}{4}$ or $x = -2$

30. $x = 2$ or $x = -2$ or $x = -\frac{3}{2}$

31. Answers may vary. A negative number raised to an even power can never be negative.

32. focus: (4, 2); directrix: $x = 2$; vertex: (3, 2); axis of symmetry: $y = 2$

33. $\frac{(x+2)^2}{16} + \frac{(y-3)^2}{36} = 1$

34. $a = \sqrt{2}$, $b = \sqrt{2}$ and $a = -\sqrt{2}$, $b = -\sqrt{2}$

35. $x = 2$, $y = 4$ and $x = -4$, $y = -2$

36. D

37. B

38. 56.3 degrees.

39. 73.7 degrees and 16.3 degrees

40. 13.2 feet from the base of the pole at a 48.6 degree angle

41. 136,080

42. 720

43. 2.5 hours

44. 50 fluid ounces

GLOSSARY

angle: Two rays sharing a common endpoint.

arithmetic sequence: A sequence in which, starting with the second term, each term is found by adding the same value, known as the common difference, to the previous term.

arithmetic series: The sum of the terms of an arithmetic sequence with a finite number of terms.

ascending order: The general practice of writing polynomials in one variable so that the exponents increase from left to right. For example, $2 + 3x - 5x^2 + 2x^3$.

asymptote lines: Dashed lines on a graph representing the limits of values where a rational function or hyperbola is defined; a graph approaches its asymptotes, but will never reach them. The exception is that polynomial graphs can cross their horizontal asymptotes.

axis of symmetry of an ellipse: Either of the two axes intersecting at its center; the longer is the major axis, the shorter, the minor axis.

axis of symmetry of a parabola: The line that passes through the vertex and focus, and is perpendicular to the directrix.

binomial: An expression containing two unlike terms separated by a + or a – sign.

binomial coefficient: The coefficient of a particular term in a binomial expansion.

binomial theorem: A formula that is used to expand a binomial to any power, thus saving a multitude of multiplications of polynomials.

center: The point in a circle from which all points are equidistant; in an ellipse, the midpoint of the segment joining the two foci, or the intersection of the axes.

circle: A conic section; the set of all points in a plane equidistant from one point.

combination: Similar to a permutation, but the order is not important. For example, the combination of eight objects taken three at a time would be $C(8, 3)$ or ${}_8C_3$.

common difference: Can be found by taking any term in an arithmetic sequence and subtracting its preceding term (see *arithmetic sequence*).

common logarithm: Understood to be base 10 when the base of a logarithm is not written (see *logarithm*).

common ratio: Found by taking any term in a geometric sequence and dividing it by its preceding term (see *geometric sequence*).

completely factored: Simplified by division, down to its prime factors.

completing the square: A technique used to solve quadratic equations.

complex conjugates: Two binomials with the same two terms but opposite signs, which represent the sum or difference of an imaginary number and a real number. For example, $a + bi$ and $a - bi$ (see *conjugates*).

complex fraction: A fraction containing one or more additional fractions (in the numerator, denominator, or both).

complex number: Any expression that is a sum of a pure imaginary number and a real number, usually in the form $a + bi$.

composite function: A function in which the variable terms have been replaced by another function.

compound inequality: A mathematical sentence with two or more inequality statements joined by "and" or "or."

conic section: A cross section formed by a plane slicing through a point-to-point pair of cones (see *circle, parabola, ellipse,* and *hyperbola*).

conjugate axis: The axis that passes through the center of the hyperbola and is perpendicular to the transverse axis (see *hyperbola*).

conjugates: Two binomials with the same two terms but opposite signs between them. For example, $5x + 3$ and $5x - 3$.

conjunction: A compound inequality that uses the word "and."

constant of proportionality: The multiplier of the independent variable in a variation relationship (usually represented by k). For example, $\frac{y}{x} = k$ (for direct variation) or $y = kx$ (for inverse variation).

coordinate plane: The x-axis, the y-axis, and all the points in their plane.

coordinates of a point: The pair of numbers in the form (x, y) designating the location of any point on a plane.

cosecant (csc): The reciprocal of sine (see *sine*).

cosine (cos): A function of an angle that is the ratio of the lengths of the adjacent leg to the hypotenuse of a right triangle.

cotangent (cot): The reciprocal of tangent (see *tangent*).

coterminal angle: Any angle that differs by one or more rotation, and whose terminal side ends at the same place as the original angle.

Cramer's Rule: A method available to solve systems of equations by using determinants; named after the Swiss mathematician Gabriel Cramer.

cross products rule: A method used to solve proportions.

dependent system: A system that contains an equation that is either a multiple of another equation or a combination of two equations.

descending order: The general practice of writing polynomials in one variable so that the exponents decrease from left to right. For example, $2x^3 - 5x^2 + 3x + 2$.

determinant: A square array of numerals or variables between vertical lines. A determinant differs from a matrix in that it has a numeric value.

difference of cubes: An expression in the form of $a^3 - b^3$.

difference of squares: A special pattern that is the result of the product of conjugates. For example, $x^2 - y^2$ is the product of conjugates $(x + y)(x - y)$.

directrix: The line from which the set of points that form a parabola are equidistant. Each point on the parabola is equidistant from the directrix and focus (see *parabola*).

direct variation: "y varies directly as x" means that as x gets larger, y also gets larger; the quotient of two amounts remains constant, $\frac{y}{x} = k$.

discriminant: The expression under the radical sign in the quadratic formula. It is used to determine the nature of the roots to a quadratic equation.

disjunction: A compound inequality that uses the word "or."

dividend: In a division problem, the number being divided into (see *quotient*).

divisor: In a division problem, the number being divided by (see *quotient*).

domain of a relation: The set of all possible x values or input values in a relation, expression, or equation.

ellipse: A conic section; the set of points in a plane such that the sum of the distances from two given points in that plane stays constant. Each of those two points is called a *focus.* The line passing through the foci is the *major axis;* its endpoints (on the ellipse) are its *major intercepts.* The line crossing the ellipse perpendicular to the major axis through the center is the *minor axis.* Its endpoints are the *minor intercepts.*

equation: A mathematical statement that says two mathematical expressions are equal.

exponential equation: An equation in which the variable appears as an exponent.

exponential function: Any function defined by $y = b^x$, where $b > 0$, $b \neq 1$, and x is a real number.

extraneous solution: A solution that does not make the original equation true. For example, the solution cannot make the denominator equal zero. It is important to check solutions and to look for extraneous solutions.

factor (n.): A number that is multiplied by another number to make a product. For example, the factors of 6 are 2 and 3, as well as 1 and 6.

factor, to (v.): To rewrite a polynomial, or a number, as a product of polynomials, polynomials and monomials, or numbers.

factorial: A way of expressing a natural number multiplied by all its preceding natural numbers. For example, 4! is read "4 factorial" and means $(4)(3)(2)(1) = 24$.

first-degree equation: Another name for a linear equation in one variable. An equation where the largest exponent of the variable is a one (see *linear equation in one variable*).

focus: A unique point in a conic section whose definition depends upon which conic section is being dealt with. In a circle, the focus (plural is *foci*) is called the center (see *parabola, hyperbola,* and *ellipse*).

formula: An algebraic equation that describes a rule, relationship, fact, or principle. For example, $I = PRT$ is the formula for finding simple interest.

function: A relation in which no domain value is repeated.

function form: The function form for the equation of a line is $f(x) = mx + b$, where the y is replaced with $f(x)$, indicating that the y value is a function of the corresponding x value.

GCF (greatest common factor): The largest expression that can be factored (divided perfectly) out of each term of another expression. For example, in $3x^2 + 6x + 12$, the GCF is 3, yielding $3(x^2 + 2x + 4)$.

general term: Each number in the sequence is called a *term*. The notation $a_1, a_2, a_3, \ldots a_n$ is used to denote the different terms in a sequence (the three dots mean to continue forward in the pattern established). The expression a_n is referred to as the general or nth term of the sequence.

geometric sequence: A sequence in which each term is found by multiplying the same value by the previous term. That value is known as the *common ratio*. Taking any term in a geometric sequence and dividing it by its preceding term yields the common ratio.

geometric series: The sum of the terms in a geometric sequence.

greatest common factor: See *GCF.*

hyperbola: A conic section; the set of all points in a plane such that the absolute value of the difference of the distances between two given points stays constant. The two given points are the *foci;* the midpoint of the segment joining the foci is the *center.* The *transverse axis* runs along the direction the hyperbola opens in. The *conjugate axis* passes through the center of the hyperbola and is perpendicular to the transverse axis. The points of intersection of the hyperbola and the transverse axis are the *vertices* of the hyperbola.

identity equation: An equation that is true for all values of the variable(s); the solution set is the set of all real numbers.

identity function: $y = x$, or $f(x) = x$; for each replacement value, the result is identical to x.

imaginary number: i represents $\sqrt{-1}$, which is an expression with no real value. It is used to evaluate square roots of negative numbers.

inconsistent system: A system of nonintersecting equations. Their solution is the null set, $\varnothing$. Their graphs are parallel.

index: In a radical expression $\left(\sqrt[n]{a}\right)$, the n, which is an integer greater than 1. If a radical expression shows no index, the index is assumed to be 2 (see *radical expression*).

index of summation: The variable defined below the Σ.

inequality: A mathematical sentence using a relational symbol other than the equal sign (=), specifically, $<$, $>$, $\leq$, or $\geq$.

inverse function:
If $f\left(g(x)\right) = g\left(f(x)\right) = x$, that is to say, when one is substituted into the other, they cancel each other out, represented by $g(x) = f^{-1}(x)$ or $f(x) = g^{-1}(x)$. Inverse functions "undo" one another.

inverse relation: The set of ordered pairs created when the x and y values of ordered pairs of the original relation are interchanged.

inverse variation: "y varies inversely as x" means that as x gets larger, y gets smaller, and as x gets smaller, y gets larger. The product of two amounts remains constant, $xy = k$.

joint variation: "y varies jointly as x and z" means that the quotient of y and the product of x and z remains constant, $\frac{y}{xz} = k$.

like radical expressions: Radical expressions with an identical index and radicand (see *radical expression*).

linear equation in one variable: An equation of the form $ax + by = c$. The graph of a linear equation is a straight line, where $a \neq 0$, and $b \neq 0$.

linear inequality: A linear sentence containing an inequality sign.

logarithm: The power to which a fixed number (the base) must be raised in order to produce a given number. Abbreviated as *log*. It is usually computed in base 10 (common logs, where the base is not written), or in base e (known as natural logs and abbreviated ln). Logs are inverses of exponents. Since $2^3 = 8$, $\log_2 8 = 3$.

logarithmic equation: An equation that involves the logarithm of an expression containing a variable.

logarithmic function: A function of the form $y = \log_b x$, where $x > 0$, $b > 0$, and $b \neq 1$.

major axis: The line passing through the foci of an ellipse, having its endpoints on the ellipse (see *ellipse*).

major intercepts: The points where the major axis of an ellipse intersect the ellipse (see *ellipse*).

matrix (pl. matrices): A rectangular array of numerals or variables that can be used to represent systems of equations.

minor axis: See *ellipse.*

minor intercepts: See *ellipse.*

monomial: A single-term expression, not containing + or – signs. For example, 5, x, $3a$, and $4x^2y^2$.

multiplication principle for events: A principle used to determine how many different ways a particular event can occur. For example, if one event can occur in p different ways and another in q different ways and p and q are independent events, then together they can occur in pq different ways.

natural logarithm: A term that represents log base e (also $\log_e$), which is written as ln (see *logarithm*).

ordered pair: Represented as (x, y). The x value is always written first, separated from the y value by a comma. The x value is the horizontal value; the y value is the vertical value on a coordinate system (see *coordinates of a point*).

origin: The point (0, 0) where the x-axis and the y-axis intersect.

parabola: A conic section; the set of points in a plane that are the same distance from a given point and a given line in that plane. The given line is called the *directrix* and the given point is called the *focus.*

Pascal's triangle: An array of numbers that are the coefficients in a binomial expansion; named after the French mathematician Blaise Pascal.

permutation: The arrangement of objects in a certain order. For example, eight objects arranged three at a time would be $P(8, 3)$ or ${}_8P_3$.

point-slope form (of a nonvertical line): Takes the following form, where $(x - x_1)$ = the difference between the x variable and the x-coordinate of one of the points, and $(y - y_1)$ = the difference between the y variable and the y-coordinate of one of the points; m is the slope. Point-slope form is represented as $y - y_1 = m(x - x_1)$.

polynomial: An expression consisting of monomial terms separated by signs.

polynomial function: Any function of the form $P(x) = a_0x^n + a_1x^{n-1} + a_2x^{n-2} + \ldots + a_{n-1}x + a_n$, where the coefficients $a_0, a_1, a_2, \ldots, a_n$ are real numbers, and n is a whole number.

proportion: An equation stating that two rational expressions are equal.

pure imaginary number: Any product of a real number and i. For example, $3i$, $5i$, and so on (see *imaginary number*).

quadrants: The four regions defined by the intersection of the x-axis and y-axis and named by Roman numerals: I, II, III, and IV.

quadratic equation: An equation of degree two.

quadratic formula: A formula that may be used to solve quadratic equations in standard quadratic form:

$$x = \frac{-b \pm \sqrt{b^2 - 4ac}}{2a}.$$

quotient: The answer to a division problem. In $10 \div 5 = 2$, 10 is the dividend, 5 is the divisor, and 2 is the quotient.

radian: The ratio between length of an arc to the length of its radius. There are 2π radians in the full revolution of a circle.

radical expression: Any expression containing a radical ($\sqrt{\ }$) symbol. For example, $\sqrt[n]{a}$ is a radical expression where the n is an integer greater than 1.

radical sign: In a radical expression $\sqrt[n]{a}$, the symbol ($\sqrt{\ }$) is called the radical sign.

radicand: In a radical expression $\sqrt[n]{a}$, the expression under the radical sign (a) is called the radicand.

radius: The distance from the center of the circle to any point on the circle is called the radius (r) of the circle.

range of a relation: The set of all the y values or output values (second number in each ordered pair) in a relation.

rational equation: An equation involving rational expressions.

rational exponent: A rational exponent is an exponent written in the form of $\frac{a}{b}$, where $b \neq 0$.

rational expression: The quotient of two polynomials, usually expressed as a fraction. The denominator of any fraction can never have a value of zero.

rational function: If $f(x)$ is a rational expression, then $y = f(x)$ is a rational function.

rationalizing the denominator: A process used to remove radicals from denominators of rational expressions.

reference angle: The term used to describe the acute angle formed by the terminal side of the angle and the x-axis.

relation: A set of ordered pairs.

secant (sec): The reciprocal of cosine (see *cosine*).

sequence: An ordered list of numbers whose consecutive terms can differ by constant, or form, in a constant ratio.

series: Sums of sequences with a finite or an infinite number of terms.

sine (sin): A function of an angle that is the ratio of the lengths of the opposite leg to the hypotenuse of a right triangle.

slope-intercept form: $y = mx + b$, where x and y are the coordinates of a point on the graph of the line, m is the slope of the line, and b is the value of the y-intercept.

slope of a line: A fraction that represents the line's rise over its run (or its change in y divided by its change in x values) as the graph of the line moves to the right. A line that descends as it moves right has a negative slope; a line that ascends as it moves right has a positive slope; a horizontal line has a slope of 0; a vertical line has no slope, or a slope that is undefined.

solution set: The set of all numbers that satisfy the given equation.

square root property: A property that is used to solve a quadratic where the square of the variable is on the left side of the equal sign, allowing us to take the square root of both sides (remembering to perform the $\pm$ on the right side).

square trinomial: The expression produced by squaring a binomial. For example, $(x + y)^2 = x^2 + 2xy + y^2$ and $(x - y)^2 = x^2 - 2xy + y^2$.

standard form (of a line): The standard form for the equation of a line; written as $Ax + By = C$, where A, B, and C are integers, A is positive, and A and B can't both be 0.

standard quadratic form: The standard form of a second-degree equation, whose graph is a parabola: $ax^2 + bx + c = 0$, where $a \neq 0$.

summation notation: Uses the Greek letter sigma (Σ) to indicate the sum of a finite number of terms in a sequence. The lower number, or *index of summation,* is the value where summation starts. The upper number is the *upper limit of summation.* For example,

$$\sum_{d=3}^{8}(3d - 8).$$

sum of cubes: An expression in the following form: $a^3 + b^3$.

synthetic division: A shortcut for dividing a polynomial by a linear binomial, where only coefficients are used.

synthetic substitution: A process that allows us to evaluate a polynomial using synthetic division. The value to the far right is the answer to the substitution.

system: A set of equations or inequalities. Solving a system means finding the values of each variable that simultaneously satisfy each equation or inequality in the system.

tangent (tan): A function of an angle that is the ratio of the opposite leg to the adjacent leg in a right triangle.

term: Any number in a sequence, or a "piece," of a polynomial separated by a + or – sign.

transverse axis: The line along the direction the hyperbola opens that passes through its vertices (see *hyperbola*).

trigonometry: The branch of mathematics that is used to find distances, angles, and indirect measurement.

trinomial: An expression containing three terms separated by + or – signs.

unit circle: A circle of radius 1, centered at the origin.

varies directly: As one quantity increases or decreases, so does another quantity (see *direct variation*).

varies inversely: As one quantity increases another decreases, and vice versa (see *inverse variation*).

varies jointly: See *joint variation.*

vertex (pl. vertices): The common point to join two line segments.

vertex (of ellipse): Either of the points on the ellipse where the ellipse intersects the major axis and minor axis.

vertex (of hyperbola): Either of the two points of intersection of the hyperbola and the transverse axis (see *hyperbola*).

vertex (of parabola): The midpoint of the perpendicular segment from the focus to the directrix (see *parabola*).

vertical line test: A test for functions. If a vertical line passes through a function at more than one point on a graph, then a domain point has been repeated, and the graphed relation is *not* a function.

***x*-axis:** The horizontal axis; all points with a y-coordinate of 0.

***x*-coordinate:** The number on the left side of the comma in an ordered pair (x, y). Also known as the *abscissa.*

***x*-intercept:** A point at which a graph crosses the x-axis. Some graphs have more than one x-intercept, and some graphs have no x-intercepts.

***y*-axis:** The vertical axis; all points with an x-coordinate of 0.

***y*-coordinate:** The number on the right side of the comma in an ordered pair (x, y). Also known as the *ordinate.*

***y*-intercept:** A point at which a graph crosses the y-axis.

zero of a function: Any value for the variable that will produce a solution of 0. Also referred to as a *solution to the equation.*

zero-product rule: A property that states if a product is zero, then either or both factors must also be zero.

INDEX

F

G

H

I

O

P

Q

R

S

Made in the USA
Monee, IL
04 November 2026

67384732R00193